Nay Shram 5/16

Thanks for
your leadership!

[signature: John G____]

Go R.E. !

The Advanced Smart Grid

Edge Power Driving Sustainability

Related Artech House Titles

Battery Management Systems for Large Lithium-Ion Battery Packs, Davide Andrea

Energy Harvesting for Autonomous Systems, Stephen Beeby and Neil White

The Advanced Smart Grid

Edge Power Driving Sustainability

Andres Carvallo

John Cooper

ARTECH
HOUSE

BOSTON | LONDON
artechhouse.com

Library of Congress Cataloging-in-Publication Data
A catalog record for this book is available from the U.S. Library of Congress.

British Library Cataloguing in Publication Data
A catalogue record for this book is available from the British Library.

Cover design by Vicki Kane

ISBN 13: 978-1-60807-127-2

© 2011 ARTECH HOUSE
685 Canton Street
Norwood, MA 02062

10 9 8 7 6 5 4 3

To Angela, Alexandra Lauren, Andres Josephe, and Austin Theodore

To Barbette, Blake, and Wesley

Contents

Foreword

Who the hell has time for vision anymore? In a world where everything is in constant motion, the luxury of a long view is a scarcer and rarer commodity than ever before. Thank goodness Andres Carvallo and John Cooper have a vision of where the smart grid is headed and the skill to succinctly describe it in this inspiring and compelling book, *The Advanced Smart Grid: Edge Power Driving Sustainability*.

For anyone and everyone in the smart grid arena, this book is an essential read. Stakeholders in the smart grid are a very large and diverse group these days. Certainly, anyone working for an energy or water utility (yes, there are smart water grids) has a big stake in smart grids, as do all the technology companies who will supply the utilities as they retool themselves into twenty-first-century utilities. (Cisco's John Chambers is so right when he says that the smart grid is far larger than the Internet.) Also, the reach of the smart grid is far greater. It encompasses the ranks of legislators and regulators who will use the smart grid to advance policies of energy independence and conservation. And, finally, the smart grid arena now embraces the consumer. With energy prices inexorably headed higher and electric vehicles coming our way, advanced smart grids will enable the tools that individuals will need to control rising energy costs.

The Advanced Smart Grid: Edge Power Driving Sustainability is an important book not only because it illuminates all the different pieces of the smart grid today, but also because it clearly shows how what has been built to date lays a solid foundation for what will become the advanced smart grid very quickly. When so many in this business seem to be hunched over a virtual microscope, examining important issues like interoperability standards and the relative value of state versus federal regulation, it is extraordinarily helpful to have Andres and John present us with this expansive vision of where we are headed. Compared to the micro focus of what consumes so much of our time today, the authors present an image of two visionaries, arms outstretched, simply saying, "Look at where we're headed. Follow us."

This book is by no means a Pollyanna-ish view of smart grids' future. It presents real issues that must be addressed, but it does so with envisioned solutions that will truly advance the discussion and avoid an all-too-prevalent tendency to circle back over issues already resolved.

The last time I enjoyed a journey without a destination in mind was a road trip to nowhere in college. As I look at the future of where we are headed with smart grids, it is great to have a map like the one that Andres and John have created.

I met Andres Carvallo when he was the chief information officer at Austin Energy at a time when he was almost finished building America's first smart grid city. When I discovered he had come into the utility industry from the IT industry (my background as well), I felt he was a kindred spirit. As I work with the Utilities Telecom Council's many utility members, it is a vision like Andres' in Austin—one that combines customer benefits, information technology management, and traditional utility communications networking—that successfully drives smart grid projects forward. Several months ago when Andres moved to Grid Net, he made it clear that the smart grid that everyone was talking about today had already been created and that it was time to move on to the next great thing. It turns out that "next great thing" is the advanced smart grid, superbly laid out in this terrific book.

Much of *The Advanced Smart Grid: Edge Power Driving Sustainability* references the superb work of the Pecan Street Project, an Austin-based research and development group working on advanced energy management systems. In 2009, the Pecan Street Project won an Energy Department stimulus grant for a smart gird demonstration project in Austin's Mueller community. In 2010, the group published recommendations from a report with well-thought-out ideas on the economic, policy, and technological implications of a power grid that relies on "better energy efficiency, locally generated renewable energy and a new economic model for electric utilities." The author of that report was John Cooper. It is on occasion joked/stated as gospel (depending on your perspective) that what is good for Texas is good for the United States. In this case, it was no joke. Many of the ideas created by the Pecan Street Project for Central Texas are excellent guidance for America and that vision comes through clearly in this book.

I was honored to be asked to write a foreword for *The Advanced Smart Grid: Edge Power Driving Sustainability* because I share the vision outlined by Andres and John. Also, I know from my personal experience with the introduction of new transformative, technology-based services—such as automated tellers and debit cards, electronic mail, and competitive communications markets—that each reaches a tipping point when the questions stop being "if" and start focusing on "when." We are at such a time with smart grids, and it is at times such as these that a strong vision of where we are headed is essential to the success of every stakeholder in the smart grid arena.

The Advanced Smart Grid: Edge Power Driving Sustainability is the visionary book on smart grids. It is the right book for these times. If you read only one book on the smart grid, you have already picked the right one.

William R. Moroney
President and Chief Executive Officer
Utilities Telecom Council
Washington, D.C.
June 2011

Foreword

In noting the technological achievements of the last hundred years, the National Academy of Engineering cited the electric grid as the greatest engineering achievement of the twentieth century. It was ranked ahead of other marvels of modern life such as the automobile, the airplane, space exploration, the telephone system, and even the Internet. Electricity is fundamental to today's society. In fact, without electricity none of the other great achievements would have been possible.

It is amazing to think that in many respects, the basic architecture of today's electric grid is little changed from the way it was designed over 100 years ago. It is long overdue for modernization. Our twenty-first-century economy needs a twenty-first-century grid to provide clean, affordable, and reliable electricity to power our society. One of the basic concepts of the smart grid is the integration of information and communications technologies (ICT) into the power system to make it more cost effective, efficient, reliable, and cleaner and provide customers with actionable information about their energy use so they can control their costs.

This is not the first time we have applied ICT to improve and transform a national or global infrastructure. In the mid-1970s, I started my career with Bell Laboratories as an architect in the Bell System's effort to completely automate the operations and maintenance of the nationwide telephone network, using distributed computing and data networking technologies of that era. At that time the computing industry was characterized by a lot of vendor-specific, proprietary architectures that did not work together. In developing systems to automate telephone network operations, we concluded early on that we had to take a different approach. Having a common, well-understood architecture and using layered protocols and open standards were critical to creating a system of systems that could evolve and adapt to incorporate new technologies, meet changing requirements, and not be locked into a single supplier. Moore's law and advances in software technology also drove much more distributed operation. Here we are, 35 years later, at the early stages of applying these concepts to the operation of the electric grid. It is long overdue.

George Santaya wrote: "Those who cannot remember the past are condemned to repeat it." A great deal has been learned over the last 35 years in the development and evolution of the telephone system, the Internet, banking networks, and early efforts to develop smart grids that need to be applied to avoid repeating the mistakes of the past. The smart grid requires the integration of knowledge about power systems and information and communications technologies, and how these technologies interact and evolve over time. The smart grid is not a static thing; like the Internet, it will evolve as experience is gained with early deployments and new technologies appear on the scene. In *The Advanced Smart Grid: Edge Power Driving Sustainability*, Andres Carvallo and John Cooper give us insights into how the evolution and integration of ICT and the electric system will power our twenty-first-century economy.

George W. Arnold, Eng.Sc.D.
National Coordinator for Smart Grid Interoperability
National Institute of Standards and Technology
U.S. Department of Commerce
Washington, D.C.
June 2011

Preface

Opportunity Meets Planning

One could say that we've been writing this book in bits and pieces for the better part of the past decade. We've certainly borrowed lessons learned from our past careers in computers, software, telecommunications, state government, management consulting, and electric utilities. The critical piece of good fortune was to have been at the right place at the right time when Austin Energy needed an overhaul of its information technology and telecommunications architecture to support its progressive vision to build the utility of the future, and then later when the community came together to launch the Pecan Street Project to build on our successful innovation at Austin Energy. But then, things tend to happen for a reason.

As no less a luminary than Thomas Edison once said, "Good fortune is what happens when opportunity meets with planning." Throughout our careers as corporate athletes and entrepreneurs, we've both looked at complexity and explored ways to simplify it, ranging from addressing complex processes and concepts in technology and engineering, to finding the kernels of truth in government research and executive briefings in the legislative and regulatory arena. Lacking precedent, we worked through the complexities of information technology, communications, and power engineering at Austin Energy identified a path to simplify and innovate to build the very first smart grid in the United States, and then refined our vision with our work in the Pecan Street Project.

This book describes in detail our experience in building the very first smart grid in the United States at Austin Energy, what we now call a first generation smart grid, or Smart Grid 1.0, and in helping to design an energy Internet at the Pecan Street Project, which we've elaborated on to create our advanced smart grid vision. In these pages, we start with that vision that sprang from those unique experiences from 2003 to 2010; then we go back to share our local perspective in Austin (it is the City of Ideas, after all); and finally spend some time sharing our observations about the past year and what lies ahead for our industry and society.

Necessarily, we focus our story on the new power engineering concepts needed to drive this transition to a more rational approach to designing and operating an advanced smart grid—look for the "Power Engineering Concept Briefs" throughout. We also included use cases where it made sense. This is a highly complex industry on a good day, and as we attempt to accomplish this fundamental transformation, it will only get more complicated. It pays to roll up your sleeves and get down in the weeds, as they say.

The remainder of this preface provides an overview of our approach, followed by a public acknowledgment of the many people who have helped us to get where we are. We hope you enjoy this book and let us know what you think. We live in an interactive world now, and this will certainly be an iterative process. Together, we'll get the smart grid right. We have to.

Chapter 1: The Inevitable Emergence of the Smart Grid

In Chapter 1, we draw a distinction between smart grids as they are described, designed and built today—what we term first generation smart grids (Smart Grid 1.0)—and second generation, or advanced smart grids (Smart Grid 2.0), which will emerge as a new understanding takes hold in this industry in line with the vision we have described in this book. First generation smart grids start with an application, such as advanced metering infrastructure (AMI), and build a smart grid incrementally by adding more applications over time. Advanced smart grids start with a smart grid architecture as part of a deliberate design that includes integrated Internet Protocol (IP) network design, thereby positioning the smart grid to support any variety of applications as necessary.

We describe the electric grid as the most important of all the infrastructures we depend upon in our modern economy and society, going so far as to insert electricity at the base of Maslow's hierarchy of needs. We assert that it is inevitable that the grid will be upgraded to become an advanced smart grid because it is the quintessential infrastructure, but also because technology evolves to empower individuals at the edge over time, and the electricity industry will follow similar trend lines described by evolution in the information technology and telecom industries.

Today, the grid is brittle and challenged, in need of a new architecture. The way forward will be through a new design and overhaul to make it more resilient and even more robust. As the number of connected devices increases dramatically, the level of complexity in the grid will rise to the point where automated protocols are needed to maintain stability—and an Internet design will be required to enable the transfer of very large amounts of data and to ensure that the grid remains functional and continues to supply us with the power on which we are so dependent every minute of every day.

Chapter 2: The Rationale for an Advanced Smart Grid

In Chapter 2 we drill down to explore the impact that extending intelligence to the edge will have on utility network architecture, business processes, and organizational structure. Distributed control systems (DCSs) have traditionally been generation-oriented, in so much as the management system was comprised of a software program running on a dedicated computer providing directions to an *automatic generation controller* at a central generation unit (i.e., power plant), to manage all the switches, boilers, and other devices through control systems, throttling the turbines up and down to maintain grid voltage levels within a specified tight band (60 Hz in the United States).

The rationale for an advanced smart grid is not hard to understand. In a sense, progress in grid management has been about gaining greater efficiencies through better control and better information. Pushing intelligence out into the grid, traditionally accomplished through independent appliances, applications, and networks, will become the purview of an integrated advanced smart grid. But for that to happen, utility business processes such as annual departmental budget building must also be addressed. The shift from an industrial approach of long product life cycles, to a more information technology (IT)-oriented environment implies not only managing a more dynamic product market, but also integration with an IT department and consideration of impacts on the greater ecosystem of interconnected devices and data.

Chapter 2: Power Engineering Concept Briefs

The advanced smart grid will require departmental managers to coordinate like never before with IT staff on security, network strategy, interconnectivity, and network integration. Common databases will drive applications, minimizing the need for complex integration projects. The benefits and implications of a system rationale will be fully explored in Chapter 2, as planning for a new era of robust digital networks gets underway, where everything connected to the smart grid has become smart in its own way. The advance of technology will inevitably encroach upon traditional utility domains, bringing changes to the traditional utility business model, and to the way a power engineer approaches grid management.

Chapter 3: Smart Convergence

Chapter 3 describes the ongoing opportunity for change that two megatrends present to every infrastructure that supports our modern ecosystem, from electricity, to telephony, to the Internet, to water, gas, and transportation. The two megatrends? First, ongoing analog-to-digital transitions are based on advances in digital technology driven by Moore's law. Digitization brings faster, cheaper, more powerful computing capabilities to edge devices that transform the potential of infrastructure design and operations. Complementing that trend is the second megatrend, advances in telecommunications and networking, driven by Metcalfe's law. As more wired and wireless technologies become available, infrastructures gain a tool to add digital devices and modernize their infrastructure operations. And as all these infrastructures begin to transform themselves, they converge on each other, offering still more synergy.

Complementing this convergence of infrastructures is a convergence of business practices from other industries onto the electric utility infrastructure, perhaps best exemplified by the addition of warehouses to the electric supply chain. Lacking a storage option, the electric utility supply chain has developed an overwhelming reliance on supply side solutions to keep voltage and VAR levels in harmony. Keeping the grid in balance is the overriding goal of the utility controller. With the addition of storage, electric utility operators will see energy storage as an alternative to generation and a valuable tool to keep the grid in balance.

Chapter 3: Power Engineering Concept Briefs

This chapter has perhaps the most comprehensive set of concept briefs of all the chapters, describing in detail how the convergence between infrastructures and the convergence of new concepts will work at the engineering level to bring changes to the grid. From a description of how to build a "thin" smart grid wireless IP network, to the detailed discussion of incorporating new digital tools into grid management solutions, these sections provide an engineering drill-down for the technically minded.

Chapter 4: Smart Grid 1.0 Emerges

Chapter 4 tells the story of how the very first comprehensive, utility-wide first generation smart grid came to be built in Austin, Texas, at Austin Energy, the city-owned electric utility that serves over 400,000 residential, commercial, and industrial customers. The lessons learned in the smart grid journey described in this case study are fundamental to understanding the concepts in this book, which derive from the successes and lessons learned in Austin, in Texas, and in the United States from 2003 to the present.

The chapter describes the initial assessment, efforts to realign IT processes and gain organizational buy-in, and communication of a new, more comprehensive vision. Key to the transition were institutional tools such as a technology governance plan, a technology leadership team, customer steering committees, a project management office (PMO), an Enterprise Architecture Council, a Technology Security Council, a Disaster Recovery Council, and an Enterprise Data Council, which together took the utility on an evolutionary journey from technical anarchy to standardization, increasing productivity and instituting proactive control.

Chapter 4: Power Engineering Concept Briefs

Any advanced smart grid project needs two things above all else. First, project funding, which can be found in part from system rationalization that eradicates wasteful IT spending and frees up cash to fund strategic initiatives like an advanced smart grid project. Second, organizational buy-in from other departments, achieved by improved communication, better service, and cross-functional experiences. In learning how to build a smart grid, the team at Austin Energy gained a valuable insight: The complexities inherent in this approach can—and should—be avoided. The best way to avoid the complexities is to start at the network level, and in particular, with a smart grid architecture framework.

Chapter 5: Envisioning and Designing Smart Grid 2.0

In Chapter 5 we tell the story of the Pecan Street Project, a unique community project in Austin, Texas, whose goal was to envision and design an energy Internet, essentially, an advanced smart grid, launched in late 2008 and continuing into 2010 as an ARRA-funded DOE Smart Grid Demonstration Project. The project was informed by the experiences at Austin Energy from 2003 to 2009 in building a pioneer first generation smart grid. The value of having weekly brainstorming over 6 months to evaluate and imagine the next generation of the smart grid proved immeasurable and helped to define the essentials of the advanced smart grid vision outlined in this book.

The Pecan Street Project, named for a main cross street in downtown Austin, began in 2008 with an idea for clean energy–led economic development. By harnessing community input, the project founders hoped to capture the imagination of the nation and steer clean energy companies to locate and grow in Austin and central Texas, creating a new focal point for economic development to complement the semiconductor and Internet business foundations of Austin's New Age economy. While the jury is still out on the success of that primary goal, the 200-person team did succeed in laying the groundwork plans for a new clean energy ecosystem, whose central tenet is the importance of integrating the water and transportation infrastructure with the power infrastructure. Beyond that level of integration, the project also emphasized the importance of integrating the community into the decision and planning process, since so much of the distributed energy resources (DER) equation depends on an informed and motivated citizenry to move beyond niche applications into mainstream adoption. Throughout this book, DER is a term that includes distributed generation, electric vehicles, and energy storage, new energy technologies that collectively comprise the supply side of "edge power." In the Pecan Street Project discussion, the term is expanded to include demand response and energy efficiency, the two demand side components of edge power.

Three key elements emerged in the storyline of Phase One of the Pecan Street Project: first, the need to integrate technology, specifically the emerging DER technologies, but also water technologies; second, the emerging need for a new business model for utilities to replace the 100-year-old model of distributing commodity kilowatt hours; and third, the need to integrate and motivate the energy consumer into the energy ecosystem. In Phase Two, the nonprofit Pecan Street Project organization is now administering DOE grant money in a 3-year study of an energy Internet neighborhood at a new urban style neighborhood located on the site of the old Mueller Airport in central Austin.

Chapter 5: Power Engineering Concept Briefs

The Pecan Street Architectural Framework, and all the other elements of the Pecan Street Project process offer tremendous lesson plans from the power engineering perspective, which is discussed in detail in this section.

Chapter 6: The National Perspective on Smart Grid

In Chapter 6 we pause from describing our local journey on the path to a smart grid in Austin at Austin Energy and the Pecan Street Project to set the context at the national level. While it remains critical to understand the changes that must take place inside a utility and within a community, it's also important to understand the context at the regional and national level. We look at the origins of the smart grid movement in the United States and globally, and then focus on three principal stakeholder groups in 2010: national and state-level governmental activity, industry standards, and industry stakeholder groups.

Chapter 7: Fast-Forward to Smart Grid 3.0

In Chapter 7 we take time to contemplate the future, in a world where the advanced smart grid has become the accepted norm and we begin to realize the system benefits of a rationalized regional smart grid. We discuss the standards and templates that are only now being contemplated, but as implemented will guide the deployment of advanced smart grids. We explore different strategies and capabilities that will be needed to overcome new complexities, challenges and obstacles that will appear on the road to an advanced smart grid.

We also examine planning and design for advanced smart grids anew, and the role of a smart grid architecture framework, which provides a cookbook of how-to recipes for grid designers. We discuss the role of a smart grid optimization engine, which provides a mechanism for planning and operation of an advanced smart grid using real-time, event-driven decisions in a data-rich environment. And we explore the integration of distributed elements, where a planner will need to one day plan and design for a future with millions of devices, far beyond what we have today.

To conclude our journey, we look briefly at Smart Grid 3.0, where a vision of a clean, linked future involves such things as peer-to-peer energy trading (P2PET), energy roaming and electric vehicles (EVs), and energy storage (ES), virtual power plants (VPPs), and microgrids. When pervasive IP networking and computing and energy become commingled with abundant information and edge-based distributed generation (DG) energy, new forms of energy trading and consumption will become possible, similar to how we currently move and consume content over the Internet.

We have termed this ambitious future Smart Grid 3.0. What comes after the advanced smart grid? We believe it is a golden age of abundance, where we manage what we have with greater respect for limits and boundaries, but we also enjoy what we have much more, thanks to sustainable networks that eliminate or minimize waste and encourage easy, even effortless transactions.

We also include use cases in Chapter 7 that describe three real-life scenarios and the different features of the advanced smart grid that will ultimately enable the Smart Grid 3.0 vision.

Acknowledgments

Beyond our families, whom we love deeply and who have been denied our company these past 2 years over innumerable weekends and long evenings while we've been busy writing this book, we'd like to acknowledge a few of the many friends, mentors, bosses, colleagues, and associates who taught us, connected the dots, were there to brainstorm with us, and, finally, inspired us with their insights over the years (and our apologies to those left off this list; there were so many more, too many to acknowledge individually): at AEP: Mike Thomason, Mike Babin, Charles Patton, and Stuart Solomon; at Austin Energy: Juan Garza, Roger Duncan, Elaine Hart, Bob Kahn, Michael McCluskey, Cheryl Mele, Chris Kirksey, Kerry Overton, Ed Clark, John Baker, David Wood, Brian Davison, Andy Perny, John Tempesta, Al Sarria, Keith Rabun, Debbie Starr, Jerry Hernandez, J. J. Guitierrez, Mike LaMarre, and Karl Popham; at Borland: Philippe Kahn, Rob Dickerson, Doug Antone, and Steve Schiro; at Cisco: Laura Ipsen, Paul DeMartini, and J. D. Stanley; at City of Austin: Will Wynn and Rudy Garza; at City of San Marcos: Mike Sturm; at DEC: Enrico Pesatori, Harry Copperman, Ada Holian, Rafael Pineiro, and Nicky Lecaroz; at Direct Energy: David Dollihite and Kelly Hayes; at Gartner: Zarko Sumic; at Dell: Dwight Moore; at GE: Bob Gilligan, Mark Hurra, Steve Richards, Kerry Evans, and Mike Carlson; at GRIDbot: Richard Donnelly; at Grid Net: Ray Bell, Stephen Street, Will Bell, and Jawed Sayed; at GridWise Alliance: Katherine Hamilton and Rich O'Neill; at IBM: Steve Mills, John Soyring, Guido Bartels, Brad Gammons, Allan Schurr, Scott Winters, Paul Williamson, Mike Francese, and Amy Thomas-Gerling; at *IDC Energy Insights*: Rick Nicholson; at iMark.com: Brian Magierski; at KU: Stuart Bell, Perry Alexander, Gary Minden, and Larry Weatherley; at Lighthouse Solar: Stan Pipkin; at Lotus: Jim Manzi and James

Henry; at Microsoft: Bill Gates, Steve Ballmer, Scott Oki, Adrian King, Greg Diaz, Gabe Newell, Cameron Myhrvold, Rob Horwitz, and Joseph Mouhanna; at NIST: George Arnold; at Northeast Energy Partners: Erich Landis; at OGE: Chris Greenwell; at Portland State: Jeff Hammarlund; at Pecan Street Project: Brewster McCracken, Colin Rowan, and Jim Marston; at Phillips Electronics: Mike McTighe, Osmo Hautenan, Terry Vega, Delia Schneider, and Paul Murdock; at SCO: Doug Michels, Xavier Montserrat, Don Morrison, Mike Foster, and David Bernstein; at Sharp Labs: Carl Mansfield and Mick Grove; at UTC: Bill Moroney; at VENewNet: Tom Dickey; at UT: Ed Anderson, Mack Grady, and Ross Baldick; at WiMAX.com: Mike Wolleben and Carl Townshend; and at Xtreme Power: Mike Breen, Carlos Coe, and Jenna Gelgand.

1

The Inevitable Emergence of the Smart Grid

Introduction

On March 5, 2004, Andres Carvallo defined smart grid as follows. "The smart grid is the integration of an electric grid, a communications network, software, and hardware to monitor, control, and manage the creation, distribution, storage and consumption of energy. The smart grid of the future will be distributed, it will be interactive, it will be self-healing, and it will communicate with every device."

He also defined an advanced smart grid as follows. "An advanced smart grid enables the seamless integration of utility infrastructure, with buildings, homes, electric vehicles, distributed generation, energy storage, and smart devices to increase grid reliability, energy efficiency, renewable energy use, and customer satisfaction, while reducing capital and operating costs."

The U.S. Department of Energy (DOE) released a handbook on the smart grid in 2009, and in the first few pages, made a distinction between a "smarter grid" and a "smart grid." By this reasoning, the former is achievable with today's technologies, while the latter is more of a vision of what will be achievable as a myriad of technologies come on line and as multiple transformations reengineer the current grid. The DOE vision for a smart grid uses these adjectives: intelligent, efficient, accommodating, motivating, opportunistic, quality-focused, resilient, and green.

In effect, all definitions of the smart grid, envision some future state with certain defined qualities. So for purposes of discussion and clarity, we have

1

adopted a convention for this book in which we refer to smart grids today as *first generation smart grids,* or *Smart Grid 1.0,* if you will. Our vision for the future we define as *second generation smart grids,* or *Smart Grid 2.0,* or as in the title of this book, we simply refer to the *advanced smart grid.* And at the end of this book, we envision a future where the smart grid has evolved to an even more advanced state, which we call *Smart Grid 3.0.*

We use a key distinguishing feature to mark the difference between smart grids as they're envisioned today and how they will evolve as experience is gained and a more expansive vision—our more expansive vision, we hope—is adopted. The difference, while it may seem trivial at first, is fundamental, and that has to do with *the starting point for the smart grid project.* If the project starts off with an *application* then that smart grid by our definition must be a Smart Grid 1.0 project. If on the other hand, the starting point is a deliberate architecture, design and integrated IP network(s) that supports any application choice, then it is a Smart Grid 2.0, or an advanced smart grid project.

Nearly all smart grid projects today start with a compelling application, whether generation automation (e.g., distributed control systems), substation automation (e.g., SCADA/EMS), distribution automation (e.g., distribution management system, outage management system, or geospatial information system), demand response, or meter automation, and then design a dedicated communication network that is capable of supporting the functionality of each stand-alone application. Evolved from the silos of the current utility ecosystem (i.e., generation, transmission, distribution, metering, and retail services), the *first generation* smart grid carries with it a significant level of complexity, often perceived as a natural aspect of a smart grid project.

In fact, a considerable amount of the complexity and cost of a first generation smart grid project derives from its application-layer orientation (Figure 1.1). Starting at Layer 7 of the OSI Stack [1], the application layer—regardless of the application—requires complex integration projects to enable grid interoperability, from the start of the smart grid project onward into the future. As additional applications and devices are added to the smart grid, whether as part of the original deployment or subsequently and over time, the evolving smart grid must be integrated to ensure system interoperability and sustained grid operations. In short, starting with the application brings greater complexity, which comes at the expense of long-term grid optimization.

The *advanced* smart grid perspective begins with a basic tenet. At its core, a smart grid transition is about managing and monitoring applications and devices by leveraging information to gain efficiency for short-term and long-term financial, environmental, and societal benefits. For a system architecture whose principal goal is to leverage information on behalf of customer outcomes, it makes better sense to start with use cases, define necessary processes, choose application requirements, optimize data management and communication

OSI MODEL

7	Data	**Application** Network Process to Application
6	Data	**Presentation** Data Representation to Encryption
5	Data	**Session** Interhost Communication
4	Segments	**Transport** End-to-End Connections and Reliability
3	Packets	**Network** Path Determination and IP (Logical Addressing)
2	Frames	**Data Link** MAC and LLC (Physical Addressing)
1	Bits	**Physical** Media, Signal and Binary Transmission

Figure 1.1   The OSI model or stack.

designs, and then make infrastructure decisions. A primary focus on the appropriate design process ensures that the system will do what it is meant to do. This key insight—*starting at the network layers rather than the application layer*—produces the appropriate architecture and design, and drives incredible benefits measured not only in hard cost benefits, but also in soft strategic and operational benefits.

Network-layer change stresses investment in a future-proof architecture and network that will be able to accomplish not only the defined goals of the present and near-term future, but also the undefined but likely expansive needs of a dynamic digital future, replete with emerging innovative applications and equipment. A well-informed design and resilient integrated IP network foundation puts the utility in a position of strength, able to choose from best-of-breed solutions as they emerge, adapting the network to new purposes and functionality, consistently driving costs out by leveraging information in new ways. The advanced smart grid is foundational; we go so far as to say that its emergence is *inevitable*.

The advanced smart grid is bound to emerge for two principal reasons. First, electricity is an essential component of modern life, without which we revert to life as it was in the mid-nineteenth century. The loss of electrical power, even for just a few hours, is the ultimate disruption to the way we live. We simply cannot live as a modern society without electricity. And second, at its core, technological progress is all about individual empowerment. But only recently have advances in component miniaturization, computers, software, networking, and device power management technology and the standards that drive their pace of innovation combined to enable individual empowerment in the electric utility industry. A new distributed grid architecture is beginning to emerge that will not only ensure future reliability, but also empower individuals in new ways.

Networks and individual empowerment define twenty-first century technology. It is inevitable that the design of advanced smart grids will begin with a network orientation that is able to accommodate any and all network devices and applications that will emerge in the future. It is also inevitable that advanced smart grids will evolve to ensure an abundant and sustainable supply of electricity and to empower individuals to manage their own production, distribution, and consumption of this essential commodity. The advanced smart grid must be robust, flexible, and adaptable, so it will be; as projects move along the learning curve, society will insist on an advanced smart grid.

The Most Fundamental Infrastructure

We require electricity to power virtually all aspects of our lives today. Electricity is used in the growth, processing, and distribution of the food we eat. Electricity is used to pump and treat the water we drink and use throughout the day. In fact, moving and treating water consumes more electricity than almost any other single function.

Considering food and water—the most basic elements that sustain us—leads one to think of Maslow's hierarchy of needs, a popular theory of human psychology introduced by Abraham Maslow in 1943 in a paper entitled "A Theory of Human Motivation" [2]. Maslow elaborated his pyramid concept more fully in his book *Motivation and Personality* published in 1954 [3]. With respect to Maslow 57 years later, we would suggest that electricity must be added to the base of his pyramid as an essential component of life, as described in Figure 1.2.

Without electricity, our modern life would grind to a halt. Maslow's hierarchy and pyramid have been challenged over the years, but they still stand as a cogent summary of what matters in life, a neat graphic on how we live our lives. By putting electricity at the base of the pyramid, we acknowledge its fundamental nature. Another way to look at the electric grid is as a foundational network, which interacts with the other vital infrastructures that support our lifestyles and economy. Consider all the other networks that we depend upon that would not operate without electricity: the *water system* that brings us fresh water, the *sewer system* that removes and processes our waste, the *natural gas pipelines* that bring us gas for household purposes like heating our homes and fueling our

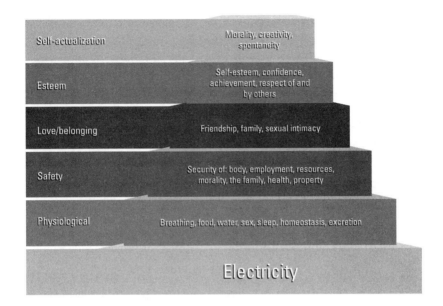

Figure 1.2 Maslow's pyramid, updated.

stoves (and to power the base load and peaking power plants that run on natural gas), and the *transportation infrastructure* with its streetlights, traffic lights, buses, and trains that depend directly upon the *electric grid* for power. The *telecommunications networks*, including wire line telephones, cellular phones, and new smart phones, would not be possible without electricity. The *entertainment media* we enjoy, from TV to radio to satellite radio requires electricity. Most recently, the *Internet* is powered by server farms and lasers shooting light beams down fiber optic lines, where electricity is so critical that massive battery banks back up data centers throughout the system in case of outages. Similarly, our *health care networks* of hospitals, medical offices, and pharmacies require electricity as a mission critical resource, and hospitals in particular rely upon ready access to backup power in a crisis.

We depend on a steady stream of electricity to our *manufacturing facilities*, and power to our *homes* to run our many household appliances, not the least of which is the humble incandescent lightbulb that started the electricity revolution. The *retail stores* where we shop use electricity for lighting, heating, cooling, and air conditioning, and to connect themselves to *financial networks* to process our purchases.

This list truly could go on and on. To drive home the point, electricity and the electric grid have become the mother of all critical infrastructures.

Ironically, the electric grid system here in the developed world has been so stable for so long that we rarely recognize that it's even there, humming away in the background, 35 feet over our heads in a ubiquitous grid of wires, poles, and towers, at least not until we suddenly experience an outage. In an instant, the lights go off, the music stops, the machines lose their spirit, give up their ghost and stop running—and then we count the seconds, minutes, and hours until power is finally restored. Is it any wonder that a blackout is so terrifying when it happens? A blackout is like hitting a giant pause button to stop our modern lives; we must wait for power to be restored before we can go on living [4]. Electricity riots, a foreign concept here in the United States where we may be accused of taking our fantastically reliable electric grid for granted, is very real around the world today [5].

The Drive to Edge Empowerment

Beyond the fact that an advanced smart grid will ensure that electricity continues to stream to millions and millions of power outlets, the advanced smart grid will arrive soon for another key reason. The principal driver of our modern economy today is the unrelenting march of technological progress pushing ever more computing and communications capability, and now energy production technology, out to the individual residential or business consumer on the edges of our networks, supplementing or even supplanting one resource or another formerly at the center of a vital distribution network. Digital technology has let the genie out of the lamp; now millions of scientists, engineers, and business people work around the clock to invent and bring to market a never-ending line of products and services based on incremental advances that empower the edge.

Ever since the integrated circuit came on line in the 1950s, the miniaturization of silicon chips and expansion of capabilities, now widely recognized by Moore's law [6], has proceeded down from micro to the nano level, putting ever more computing power on an ever-shrinking piece of real estate. Complementing the march of technological progress in computing power has been a steady advance in communications technology, with fiber optic technology revolutionizing the wired world and wireless advances creating a family of options for sending and receiving radio signals. And the Internet has extended itself to every corner of the world in remarkable speed in less than two decades, going through several phases, morphing into an ever more powerful force for delivering technological empowerment to the edge.

Beyond Moore's law, Metcalfe's law [7] comes into play at the telecommunications and Internet levels, suggesting that the value of a network is proportional to the square of the number of nodes on that network. As more and more devices are added, the network becomes ever more valuable because the number of connections goes up so rapidly. While the exact value proposition of Metcalfe's law has been subsequently challenged in light of the dot-com bust 8 years ago [8], the gist is that networking adds tremendous value (we'll keep citing Metcalfe's law as shorthand for the value proposition of networking).

To understand the impact that Moore's law and Metcalfe's law have in value expansion and on potential development milestones for the electric industry, let us compare development milestones in the information technology (IT) and telecommunications industries to what we project for the electric industry, as shown in Figure 1.3. Mainframes and central switches were not eliminated when distributed options came online, but the focus did shift to the new distributed edge, where greater computing power and communication flexibility enabled laptops and cell phones. Similarly, it is not hard to imagine today's massive central power plants gradually losing their dominance as more and more distributed generation comes online. The reason these trend lines are so similar is that they are all based on fundamental aspects of digital and network maturity: increasing computation and mobility lead to greater enablement at the edge.

The Roots of Smart Grid

Today's smart grid projects have their roots in converging streams of innovation over the past two decades, which in turn built upon historic innovation from

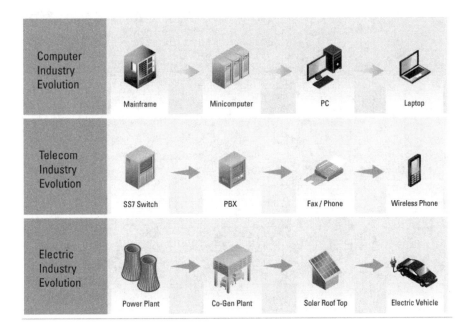

Figure 1.3 The drive to edge empowerment.

about the 1950s on (Figure 1.4). Rudimentary supervisory control in the 1950s used pilot wire systems, tonal systems, and twisted pair copper wires. Likewise, early telemetry was employed to provide power readings on remote elements of the grid using multiple communication media, including pilot wire, power line carrier, and microwave systems. The growth in computing capacity and solid state systems led to supervisory control and data acquisition (SCADA) systems in the 1960s.

At the ends of the SCADA systems within utility substations were remote terminal units (RTUs) connected to batteries, which enabled energy control center operators to have visibility and control of multiple elements of the grid within a substation. Inside the control center, map boards that displayed the one-line diagrams of the system were complemented (and eventually replaced) by CRT screens and display software that evolved to provide greater visibility and control. Gradually, as system complexity grew and the grid expanded, humans ceded management functions to increasingly automated systems in order to promote more efficient central control, thanks to emerging technology.

In this way, advances in telecommunication and information technology paralleled and complemented advances in power systems. Distribution and substation operations became more and more automated as the system grew in complexity. *Distribution automation* holds greater promise still going forward as the core component element of an advanced smart grid system, where

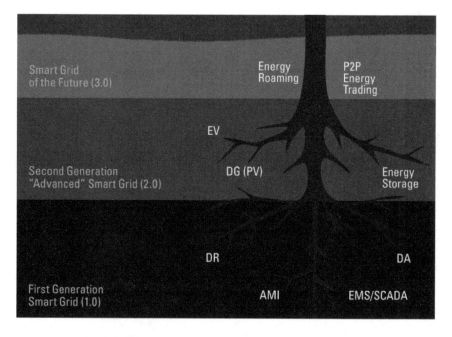

Figure 1.4   Roots of smart grid.

integration is the next logical step in this long evolution for greater control and efficiency.

And yet, throughout this long history of evolving complexity and greater use of technology, out at the very ends of the distribution system the analog meter remained, its internal gears driven by electric current flowing through the meter, with corresponding dials under the glass counting off the kilowatt hours, there to be read manually by the roaming meter reader once each month. Solid-state technology employed in multiphase meters came to commercial accounts long before new technology emerged on the residential side, where good enough was sufficient to provide the minimal requirement to produce a bill, a monthly read on kWh consumption.

However, by the 1980s, technology came to the residential meter with drive-by automated meter reading (AMR) as offered by Itron [9], where equipment in a trolling van received short-range narrowband radio signals to read new digital meters far more efficiently than a team of walking meter readers could. Similarly, the TWACS (DCSI, now Aclara [10]) and Turtle (Hunt, now Landis+Gyr) [11] products were early efforts to read meters remotely over narrowband power line carrier (PLC) technology, which became popular in less densely populated rural districts. Early AMR was sold principally as a labor-saving solution, replacing the meter reader but still producing a once-a-month read to generate a monthly electricity bill.

In the mid-1990s, Cellnet (now Landis+Gyr) [11] came on the scene offering a revolutionary fixed wireless system (i.e., RF mesh), where meters transmitted to collectors located in contiguous cells that collected data and sent them in regular bursts to still larger cell collectors for wired transmission back to a central hub. For the first time, interval data was possible to provide a more detailed picture of consumption than a monthly read, but corresponding billing systems would have to catch up in order to take advantage of all that meter data. In time, narrowband RF mesh advanced meter infrastructure (AMI) vendors proliferated, complemented by the continued use of power line carrier systems for more remote service territories.

Electricity and Telecommunications

Utilities have had a back-and-forth relationship with telecom providers over the years. The build-or-buy discussion has proceeded in every utility, as managers weighed the relative merits of investment in communications infrastructure versus outsourcing. Advances in fiber optic technology in the early 1990s proved a boon to utilities, which took advantage of such advances to create new utility telecom divisions and subsidiaries. PLC blossomed into broadband over power line (BPL) and millions of dollars were spent to determine if the power line assets of utilities could be repurposed as communication assets (see Current Group [12]). From proprietary microwave technology, wireless technology progressed in the early years of this decade to open standards based Wi-Fi, WiMAX, and ZigBee. And this book will continue the debate over the relative merits of narrowband versus broadband and mesh versus point-to-multipoint network architecture.

Handheld radios, digital pagers, and later cellular phones and wireless laptops kept line crews in touch with central dispatchers. Similarly, technology found its way into utility offices over the years, paralleling the growth of office technology in other sectors. Billing systems, accounting systems, work management systems, telephone systems, desktops, and servers and data storage grew under utility IT department supervision. As utilities moved from mainframes to minicomputers to PC systems, they grew accustomed to connecting with the outside world in new ways, but moving beyond the billing envelope and the telephone as the primary means of communication with their ratepayers proved a challenge to utilities. Gradually, utility Web sites proliferated and today, new technologies like Twitter and Facebook offer utilities new tools to interface with their external stakeholders.

If this short review of utility technology shows anything, it shows an already complex industry adapting to ever-increasing complex demands by incorporating new tools and technologies as they became available. The IT, telecommunications, and electricity sectors grew hand in hand. The electricity from the grid was vital to feed the IT and telecom revolutions. Likewise, IT and telecom advances have proven equally fundamental to the management of an increasingly complex electricity grid. By the first years of the new century, the deep thinkers in this industry who contemplated the long-term future of the grid began to conclude that increasing complexity and continuous technological improvement would lead to ever greater intelligence and automation in the grid.

Defining Smart Grid

The Electric Power Research Institute (EPRI) has promoted the *Intelligrid* concept for nearly a decade [13]. For nearly that long, IBM has been consistent in its effort to draw attention to the need for an *intelligent utility network* [14]. Rick Nicholson, today a well-known industry analyst at *IDC's Energy Insights* [15], suggested in late 2003 while at the Meta Group that the emergence of a new electricity grid architecture would ideally parallel that of the Internet, which he termed an *electric geodesic network*. Carnegie Mellon, now the home to the *Smart Grid Maturity Model* [16] initiated by IBM and others, is also a leader in R&D regarding the ultralarge-scale systems. At individual utilities like San Diego Gas and Electric, Southern California Edison, and Sacramento

Municipal Utility District, experimentation has proceeded apace. At Austin Energy, smart grid innovation has recently expanded to integrate the electric grid with buildings, homes, and transportation infrastructure. Like a family on holiday gathered around a dining room table working on a jigsaw puzzle, leading thinkers have pored over new combinations of technology, sowing the seeds for today's discussion about the smart grid.

Two major themes stand out in this brief review of the roots of today's smart grid evolution. First, utilities are traditionally organized into silos, from generation to wholesale operations to transmission to distribution to retail operations. These distinct business lines depend upon each other, but operate with great independence. Consequently, it's not surprising that the vendor community has evolved over more than half a century to meet specific needs in each of these silos. A few very large companies like GE, Siemens, ABB, IBM, and Oracle provide services across the silos, but an army of specialized vendors provide services to single silos as well, meeting the needs of demanding end users with compelling problems to solve. Progress has been measured by innovative individual applications that do more things more efficiently. The idea of a new network approach to solve this ever-expanding complexity is novel, but growing, which leads us to the predominant theme of this book.

Design: The Twenty-First-Century Smart Grid Challenge

Today, our grid needs more than a facelift, it needs to be redesigned. Today's electricity grid was designed using nineteenth- and twentieth-century sensibilities, under the limitations of the technology from a bygone era. Few would challenge the tremendous positive impact that the modern electric grid has had on the lives and economies of the residential and business customers who have enjoyed reliable electric power drawn down from the grid. But the grid—indeed, the entire electric utility industry and business model—is increasingly challenged by the need to provide ever more power at better quality yet reduce costs, by the need to accommodate new and innovative but highly disruptive technologies that in many cases actually reduce revenues, and by the need to reduce an historic, overwhelming reliance on fossil fuels. The only way to address these compelling challenges and conflicting priorities and achieve sustainable long-term success is an overhaul of both the architecture of the current grid and the business model that relies on commodity kilowatt-per-hour delivery. Answering such questions as: How did we get here? Where do we begin? Where should we go? and of course, Why go through with this? is the reason for this book.

In contrast, consider today's telecommunications network, which was originally developed about the same time as the electric grid, and under similar technology conditions. While the evolution of the telecom network paralleled the evolution of the grid, telecom development differed by the presence of a huge monopoly. That huge monopoly went through a more radical redesign, first with the breakup of ATT in 1984 [17], then with the incremental addition of wireless services starting in the mid-1980s, the Federal Telecom Act of 1996 [18] and then with the more gradual but more radical emergent Internet. Like the electric grid, the telecom network was not immune to significant change, which had started much earlier. Midway through the twentieth century, digital innovation from Bell Labs radically transformed the potential of telecommunications, bringing about digital switching, fiber optics, and laser technology to enable ever greater efficiency, capacity, and speed. However, the radical redesign of the network architecture that would ultimately take telecommunications to an altogether new level required three major milestones: first, a decision by the federal government to break the dominant monopoly into pieces to engender competition; second, the birth of cellular wireless to foster a mobile focus; and third, the emergence of the Internet as a fundamental information infrastructure to reshape the modern economy.

Arguably, the electricity grid has not faced such pressure for radical redesign yet, but we're getting there. To draw the contrast, the Department of Energy (DOE) introductory *Report on Smart Grids* [19] tells a story about Alexander Graham Bell and Thomas Edison, contrasting changes in telecom and electricity. Were Alexander Graham Bell to return today, the story goes, he would not recognize the telecom network elements in quite the same way that Thomas Edison, on a similar journey into the future, would recognize an electric utility's key components. To be sure, the electricity grid has not stood still—it continues to evolve to meet ever greater demand and to address short comings and risks, which range from the truly rare massive blackout to the far more frequent and numerous minor outages. However, the grid's historic design constraints, such as a reliance on just-in-time production to meet demand and the need for ever more delicate balancing of grid voltage put the grid at increasing risk of disruption and limit its economic potential as complexity increases and new threats emerge.

In fact, the modern electric grid has long been a model of reliability, especially considering its sheer scale and complexity. Given the critical nature of the grid and the need for reliability, the potential risk and cost of significant change, the historic conservative nature predominate in the utility culture, the lack of leadership at the federal level, and the unprecedented nature of a grid redesign, it is no surprise that any call for such dramatic change would see a slow, even skeptical response. However, the potential for a solution is here today, and this book states unequivocally that these changes must happen, and demonstrates the power engineering concepts needed to light the way forward, putting our essential, foundational network back on more solid ground. The electric industry has yet to fully embrace the recent lessons learned from the Internet, which provides not only new ways to operate at the granular level of tools and applications, but also an entirely new architectural model to accommodate the emerging needs of this new century.

Nature and the Internet: Models for Organizing Complexity

The human body is a shining example of the way that networks organize complex systems. Each of the systems inside the human body can be seen as a highly adaptable network, more accurately *nested networks*, working together under the central control of the brain, but also replete with autonomous behavior apart from central, top-down control. When a person's hand touches a hot stove, for instance, by reflex it draws away immediately based on preprogrammed intelligence that resides in a different part of the brain away from conscious thought, as a matter of survival. No conscious decision is made to withdraw the hand from the heat; it was reflex reaction to the heat. In a similar way, the advanced smart grid will see autonomous behavior from preprogrammed control messages that go out to the edge. The future of the electric grid lies down this path: to mimic the architecture of both the Internet and natural networks and systems to enable sustainability and provide the ultimate in adaptability.

From the outset, the Internet was intentionally designed to be a network solution to provide greater resiliency and to be able to adapt to ever greater complexity. Its elegant architecture has allowed it to grow rapidly and adapt to new applications, new uses, and a dramatic expansion of traffic. In turn, the Internet has taught the world valuable lessons on the way that networks work, and provided great insights into how the natural world works as well. Thanks to the Internet, we have also come to understand how the incredible complexity of the natural world is accommodated not just through the trial and error of evolution, but also through the use of self-healing networks that organize and adapt to ever-changing individual network elements that combine to form a sustainable ecosystem.

Digitization introduced the grid to the tremendous potential for change in the 1960s, and decades later, the Internet has doubled down on that promise, adding the potential of ubiquitous, high-speed connectivity as broadband technologies gain widespread adoption. Digital technologies have been implemented by electric utilities throughout the supply chain over the years. Both in the corporate offices and out in the field, utility managers worked with vendors to replace their analog devices and solutions with digital devices and applications. However, out in the field, special wireless networks accompanied digitization and soon utility telecom personnel found themselves supporting multiple, incompatible networks. Such a situation was not intentional, but came about because utilities, like most large organizations, organized themselves into independent silos.

So it has happened that utilities adopted Internet architecture in their office environments like any other large organizations, but their field operations ended up evolving down a different path. Applications customized to meet specific needs now prevail out in the field, and more often than not, the supporting networks for these applications are not IP capable. Vendors have historically offered specialized networks, not only to optimize technological performance, but also to make it hard for competitors to gain traction in their accounts. Understandably, this works out well for vendors, but utilities pay a price when they are left to support multiple networks and are locked in to a specific technology and a specific vendor. This book will show that IP has become so dominant, and that security protocols have developed, such that the time has come to take full advantage of all that IP and Internet architecture have to offer, both in the office and out in the field. The time has come for an integrated, IP networking environment in the electric utility grid, as evidenced by the themes outlined next.

The Inevitable Themes of Change

One may well ask why it is *inevitable* for the electric grid to evolve into a new architecture that more closely resembles the Internet and the networks in nature, as the title of this chapter proclaims. Is anything truly inevitable? In this case, yes, the transition to a new architecture is inevitable because the geodesic Web design is superior for flexibility and adaptability in a highly dynamic and unpredictable environment and the current design, as this book will show, is neither suitable for the evolving nature of the grid nor sustainable over time. It may be argued as to when the transformation will take place, but take place it must. There is too much at stake, and there are no suitable alternatives.

A handful of principal themes work together to drive the discussion on why a new grid architecture, an advanced smart grid, is not only needed, but inevitable. We will show in detail how we can go about getting such a grid. The advanced smart grid is not only inevitable—it is *available* today. But to fully grasp the need for new grid architecture, it is vital to understand the significance of the following change themes and what they imply.

Smart Devices and Ubiquitous Connectivity

First, *devices and connectivity* define the state of the grid—what is connected to the grid and what is not at any given time, and how many connections there are. The emergence of demand response (DR) and distributed energy resources (DER), a new class of edge power devices and systems that includes distributed generation (DG), electric vehicles (EV), and energy storage (ES), means that the industry will need to adapt from connecting a relatively manageable number of devices to connecting a *massive number of devices* under conditions of far less control. Also, entirely new processes must be incorporated into grid operations. For example, while the relatively few central generation resources are rarely disconnected today (operators ramp them up and down as needed to maintain grid voltage targets), the far more numerous DER devices, including tens of thousands of small generation plants (e.g., rooftop solar PVs, electric vehicles), will in fact need to be connected and disconnected frequently to preserve grid stability. As DER devices gain traction, they will introduce a need for utility managers to promote *resource islanding* as a strategy to maintain grid stability (resource islanding is described in greater detail in Chapter 7). And as the number of connected devices increases dramatically, the level of complexity in the grid will rise to the point where automated protocols are needed to maintain stability, and an Internet design will be required to enable the transfer of copious amounts of data and to ensure that the grid remains functional and continues to supply with the power we are so dependent on.

In the traditional electric utility, a relative handful of generation resources supplied power over the grid to first thousands, millions, tens of millions, hundreds of millions, and then billions of energy-consuming devices. Standards in grid design and standards in operating protocol, as well as standards in the appliances, switches, and plugs on the ends of the network outlined by organizations like the Underwriters Laboratory [20] ensured grid stability and harmony. Even as the number of energy-consuming devices multiplied over the latter half of the twentieth century, maintaining adequate voltage levels was managed by making generators ever larger by adding specialized peaking units and by building more distribution substations to accommodate growth. The grid was able to maintain harmony because the change so far had really been along only one dimension—adding more load—and the solution remained relatively straightforward, if expensive: add more central generation and beef up grid capacity.

On the other hand, the change driven by technological advances emerging today and the technology that waits on the horizon will be on *multiple* dimensions. First, expansion of population and adoption of consumer electronic devices and new appliances means that the number of smart devices and appliances, especially digital appliances, is accelerating the pace of change on the load side. Although if that were all, it would be manageable, as more load is merely an acceleration of the trend of load growth we have grown used to. But a new dimension of change has opened up when *distributed generation* devices like rooftop solar photovoltaic, microgas turbines (using combined heat and power), and microwind, to name just a few, put power generation control out on the edge as well as back at the core.

Distributed generation presents four key differences: (1) utility managers lack the fine control over these new generation devices (assuming that many of them are owned by utility customers), (2) the number of devices is dramatically increased at an unpredictable rate, even though the size of each generator is relatively miniscule, (3) renewable energy generators that depend on the sun or wind do not produce predictable steady streams of power, but unpredictable variable power, and, finally, (4) the power is located at the opposite end of the grid, near the load it serves, introducing a new, revolutionary issue—two-way power flow and the need for improved equipment protection. The dawning era of advanced smart grids and DER will see millions to tens of millions of new grid-connected devices attached to each grid. The management challenge will not just be difficult in the future, it will be impossible without significant changes.

Static Vesus Dynamic Change

Change—both the pace of change and the approach to change—is a critical management issue in the electric industry. The utility environment has traditionally changed only gradually, and under a controlled setting for the most part, a condition that could historically be described as *static*. As the transition to a more *dynamic* state occurs, change is becoming more frequent, less predictable, and increasingly, out of the direct control of utility managers. The utility approach to change management (cultural and organizational) will need to adapt to the new more dynamic state. Consider, for instance, that the current approach to change is based on a specific purpose (e.g., meters, distribution automation, demand response), which retains the silo focus so typical of utilities. In contrast to such purpose-driven change, *service-driven change* (e.g., service-oriented architecture or SOA) adopts a more holistic, long-view that ensures that all parts of the whole function well together by design, by more fully leveraging a ubiquitous, connected environment.

Consider the pace of change in other industries, specifically the IT industry. Driven by increasing value based on Moore's law and Metcalfe's law, change has become a constant in the IT world, and rapid obsolescence is assumed and built into the product life cycle. As more and more technology creeps into the electric utility industry and as it begins to use more and more IT and telecommunications, is it unusual that the pressure to adapt and change more rapidly would find its way in? When analog meters that had lasted more than 50 years are replaced by digital smart meters, even with strategies that extend product life to 15 years, what is to happen? The utility will be forced to adapt to a new time cycle, where change happens much faster. It will need a new attitude about change, a new approach.

In a static environment with little and/or infrequent change, it made sense that purpose-driven change would be the norm. In other words, one only changed from the status quo when a specific purpose required change, suggesting a specific application such as distribution automation or demand response. But a more highly dynamic environment, where change is far greater and occurs more frequently (the technological innovation wave reflected by DER alone promises to be highly disruptive for utilities) will require utilities to adopt a new attitude for change, namely, to prepare and acquire the necessary skill sets to make adaptability a core competency. Utilities must adapt to this shift at the organizational level by transitioning away from traditional silos for more interoperability and cross-training, they must adapt at the economic level by shifting their business models to be more oriented towards services and less towards sales of commodity kilowatts-per-hour, and they must adapt at the technology level with a new IP network architecture that will more flexibly accommodate new technologies.

Innovative Design as Change Agent

Innovation at the network level will be led by *network redesign* as the foundational theme, the adaptation to an exponential increase in the number of connected devices, to new stability mechanisms like *resource islanding*, to the two-way flow of power and information on the grid, and to a new state of permanent change where everything is in flux. But control will be maintained by automated policies and protocols out at the edge, by digital devices and intelligence spread throughout the grid all feeding data back to power processors and huge, shared databases, with complete visibility for human controllers there to operate the network with state-of-the-art tools. The new grid architecture itself will have been reinvented from its traditional radial design with relatively predictable, one-way power flow from a few generators out to passive, dumb loads on the edge, to a *Web* design with highly dynamic, unpredictable two-way power and information flow from hundreds of thousands of generators and storage units sitting alongside intelligent, active loads that also participate in keeping the grid in harmony.

Today an electric grid design is dominated by two main design types: first, a network design, using N-1 redundancy [21] for reliability flow. The network design is most often found in more densely populated urban areas. Second, a radial design, which resembles fingers extending out from the palm of the hand, is more common in suburban or rural populated areas. Current network approaches may be suitable to manage an advanced smart grid, but they might be not affordable. The radial design will certainly not be suitable for two-way power flow. So we need a new design, one that is hybrid and affordable and like the Internet will support multiple nodes connected to each other via nested networks with distributed intelligence on every device. We will cover this new design and its merits versus the current available designs in Chapters 2 and 3.

On the supply side, early pioneers started making electricity with small coal-fired generators, but soon the focus shifted to water-driven turbines (hydro) and then back to ever larger coal-fired generators. In time, supply-side innovation added natural gas as a fuel source, then nuclear and further innovation led to new gas-fired peaking units, first simple cycle, then combined cycle. More recently, renewable energy sources like wind and solar power have emerged to garner our attention. Likewise, demand-side innovation focused first on the incandescent lightbulb then went on to a variety of electric industrial and household appliances, from powering commercial ice houses and electrified automobile plants to the electric iron and refrigerator. Innovation on the distribution grid, in between production and consumption, responded to the innovation at the edges, but was driven by silo applications. With an advanced smart grid, innovation by utility managers will be driven by the network. While innovation at the edges will continue, even increase, the emerging new architecture will bring a renewed, more intense focus on innovation at the network level. In the future, innovative network design will drive innovation at the edges of the network.

New approaches and ways of thinking about systems are already emerging in the smart grid arena. In the distribution automation category, exciting work is underway to integrate three systems that currently operate independently: the Geospatial Information System (GIS), which tracks utility assets out in the field; outage management systems (OMSs), which reactively respond by correlating data inputs during an outage to route those assets to where they are needed; and distribution management systems (DMSs), which are being developed to automate many distribution functions and proactively respond to outage information while integrating all the features of a traditional OMS and enabling dynamic GIS evolution. Advanced smart grids require focusing on such key energy concepts as power quality, modulation, harmonics, and fault detection. More information gathered from more places around the grid brings great promise to bring better grid management into the distribution system, improving both routine and exceptional grid management scenarios.

Conclusion

The advanced smart grid is a concept for today, to provide us all with a vision for what lies beyond the considerable work in the trenches now underway in the United States and around the world. Most of the perspective in this book is in fact taken directly from the authors' experiences in the United States, although we acknowledge from time to time in this book the considerable work and success on smart grid outside the United States. We provide a framework and a vision that we hope will stimulate the national debate around significant issues that tend to get overlooked when focused on solving immediate, confusing and complex problems. The smart grid is one of the most vexing challenges humans face today. Upgrading our foundational electric grid while maintaining reliability, overhauling our relationship and our understanding of the nature of energy, cooperating to invent new regulatory, economic, and legal mechanisms and institutions—all these tasks require an inspiration beyond the necessity to maintain focus and drive to reach our short term objectives, as important as they are. Herein, as our contribution to the national conversation now underway in earnest on this vital topic, we offer our experience and vision for a future of advanced smart grids that will enable the widespread adoption of edge power devices and systems and ensure a sustainable platform.

Endnotes

[1] The Open Systems Interconnection (OSI) Model, often referred to simply as the "OSI Stack," is a model developed by the International Standards Organization (ISO) to explain the function ality of a communication system into layers that interoperate in a logical way, by providing unique servers up or down to another layer. The stack starts with Layer 1 and progresses to Layer 7. In Chapter 4, this book presents the fundamental argument that starting a design process after an application choice has been made—starting at Layer 7, in other words—limits the design, whereas starting an architecture design before the choice on applications, avoids such constraints and ensures that the architecture will provide a design that enables the desired functionality.

[2] "A Theory of Human Motivation," A. H. Maslow, originally published in *Psychological Review,* Vol. 50, 1943, pp. 370–396, http://psychclassics.yorku.ca/Maslow/motivation.htm.

[3] *Motivation and Personality,* A. H. Maslow (1954), http://www.abraham-maslow.com/m_motivation/Motivation-and-Personality.asp.

[4] *The City of Ember,* a post-apocalyptic novel by Jeanne DuPrau, made into a movie with Bill Murray, captures the terror of a failing electricity infrastructure; see http://en.wikipedia.org/wiki/The_City_of_Ember.

[5] http://www.google.com/search?q=electricity+riots&rls=com.microsoft:en-us:IE-SearchBox&ie=UTF-8&oe=UTF-8&sourceid=ie7&rlz=1I7ADFA_en.

[6] Moore's Law, http://www.intel.com/technology/mooreslaw/.

[7] http://en.wikipedia.org/wiki/Metcalfe's_law.

[8] http://spectrum.ieee.org/computing/networks/metcalfes-law-is-wrong/0.

[9] http://news.itron.com/Pages/ami4_0808.aspx.

[10] http://www.aclara.com/Pages/default.aspx.

[11] http://www.landisgyr.com/na//en/pub/index.cfm.

[12] http://www.currentgroup.com/.

[13] http://intelligrid.epri.com/.

[14] http://www.ibm.com/smarterplanet/us/en/smart_grid/nextsteps/solution/
      L420447J94627B46.html.

[15] http://www.idc-ei.com/.

[16] http://www.sei.cmu.edu/smartgrid/.

[17] http://en.wikipedia.org/wiki/Bell_System_divestiture.

[18] http://www.fcc.gov/telecom.html.

[19] http://www.oe.energy.gov/DocumentsandMedia/DOE_SG_Book_Single_Pages(1).pdf.

[20] http://www.ul.com/global/eng/pages/.

[21] http://en.wikipedia.org/wiki/N%2B1_redundancy.

2

The Rationale for an Advanced Smart Grid

In Chapter 1, we drew a distinction between the conventional view of smart grids and something altogether new, an advanced smart grid, as we introduced a somewhat radical concept. It is *inevitable* that the advanced smart grid will emerge. The transformation coming to the electric industry as technological innovation crashes over the utility landscape like a giant wave will leave dramatic change in its wake. However, there is more going on here than meets the eye. The role of the utility IT department is transitioning and becoming far more strategic than it has ever been before, as utilities adopt digital devices and shift from the long product life cycles of industrial, electromechanical equipment to the more dynamic, much shorter IT product life cycle, and as digital equipment in the field is networked into an energy ecosystem of interconnected devices, massive amounts of new data will need to be processed, stored, and accessed to and from common servers and databases. The shift from vertical silos to a horizontal energy ecosystem will ensure that the IT department in utilities becomes a strategic function.

Introduction

Adapting to these changes will sorely challenge a staid utility organizational culture, but it will also lead utilities to solve challenges that have long bedeviled them, from how to flatten the utility load curve to how to reduce line losses, from how to improve customer service to how to lower the utility carbon footprint. As change progresses, the benefits of change will become ever more apparent and utilities throughout the industry will become more concerned with how to adapt to change than they will with whether change was even a good idea to begin with.

Understanding the "why" of change lays the foundation for understanding the "how" of coping with the challenges that change brings. So here in this chapter, we dig deeper to explore the rationale for an advanced smart grid. First, we'll describe how such transformative changes lead to the emergence of new energy architecture to meet the needs of a new energy economy. Such changes will inevitably challenge existing business practices and processes and ultimately demand a new set of rules, assumptions and organizing principles. The changes will take years to accomplish, given the foundational role the grid plays in our lives and our economy, so we will have time to work through the complex details to get to the other side.

We'll explore why a network orientation drives the subsystems within a network to become more integrated with each other and how that must happen. Finally, we'll discuss the planning elements required if such an advanced smart grid vision is ever to be achieved.

A New Set of Rules and Assumptions

The new set of rules, assumptions, and business practices will develop in three principal areas, each building successively on its predecessor: security, standardization, and integration. *Security* has been a key focus of the National Institute of Standards and Technology (NIST) Security Plan [1], issued in August 2010, as well as the NIST Smart Grid Interoperability Panel (SGIP) [2], spelling out a baseline framework for development of *interoperability standards* that will guide industry as it creates the new smart grid economy. The reality of a networked economy is that all network components affect each other with varying levels of antagonism, constraint, and synergy. Thus, *integration* will become more and more critical as it becomes increasingly apparent that the old silo approaches that previously separated parts of the utility from each other for improved functionality now inhibit adoption of new network efficiencies.

Security

The growth of the advanced smart grid depends upon the development of security technology. As the foundational infrastructure, the smart grid cannot afford to get out in front of its ability to remain secure. Economic performance, as well as public and personal health, safety, and welfare, all depend on maintaining a secure and reliable electricity distribution network. The challenge will be to ensure that the grid remains secure as a resilient, mission-critical network capable of providing reliable, affordable power, even as it transforms into a new, very different infrastructure with capacity for so much more.

While the utility industry has seen considerable progress towards ensuring smart grid security, it has also acknowledged the considerable challenges that remain, even as it works to develop a more robust, standards-based approach. Security is unavoidably a dynamic challenge, where systems built to defend against a range of threats must evolve as the threats themselves adapt and change. In a world of dynamic threats, there is no place for a static security solution. NIST published a comprehensive list of cyber security standards in August 2010, to ensure security standards that will support the pursuit of grid interoperability [3].

NIST leadership on security is welcome news, for without the NIST standards, strategy, and framework effort, utilities would find it nearly impossible to hold smart grid technology vendors accountable. But forging security standards is far from easy. The U.S. Government Accounting Office (GAO) challenged NIST findings on smart grid security in January 2010 [4], and in late January 2011, NIST Standards were challenged again when presented to the FERC, followed by an announcement of a three-way collaboration on smart grid security by the U.S. DOE, NIST, and the North American Electric Reliability Corporation (NERC) [5]. The challenge before the industry is to ensure that the promotion and advancement of cyber security becomes an integral element of grid transformation. Smart grid solution providers must implement security identity, encryption algorithms, security protocols, and crypto key management systems that are open and standards-based, robust, proven, scalable, and extensible. Standards-based security must be designed and deployed into every aspect of the smart grid, supporting governmental and regulatory cyber security principles of confidentiality, integrity, availability, identification, authentication, nonrepudiation, access controls, accounting, and auditing.

For the advanced smart grid to be truly secure, security standards must meet the following four minimum requirements:

1. *Granularity at the device level.* Security standards must provide for the identification and isolation of compromised or hacked devices to prevent damage from spreading unchecked through the network.

2. *Standards-based security.* Security standards must be based on best-in-class protocols and requirements that leverage worldwide efforts to develop faster, simpler upgrades to produce sustainable end-to-end security

3. *Multilayer, multilevel security.* Security standards must ensure multiple safeguards in edge devices, embedded applications, network infra-

structure, network operating systems, data, and utility enterprise systems

4. *Sustainable security.* Security standards must sustain investment in security oversight, software upgrades, and process improvements and be capable of routine, automatic security updates.

Standardization

We've learned a considerable amount about the value of standards over the past two centuries. In the 1800s, it was common for an artisan or craftsman to work alone and make unique goods. Over time, the business world has slowly embraced the efficiencies and benefits of reaching economies of scale through standards achieved based on a foundation of common protocols, interfaces, and form factors that allow industry stakeholders to evolve their products in a common fashion, differentiating their products by augmenting a standardized product through innovation.

Leaders in the technology era have built on progress in the industrial age using an explicit, strategic focus on standards, gathering together corporate representatives to forge industry agreement on design, production, and operational guidelines that allow products to interoperate efficiently to provide consumer benefits and stimulate demand. Such *industry standardization* moves the whole industry rapidly through the product adoption cycle and lowers the costs of production as cost efficiencies are achieved.

Standards played a role at the national level in the early days of electricity. Household appliances benefited from industry agreement on common electricity operating designs and UL standards on plugs and switches. Likewise, the PC and consumer electronics industries have benefited from standards, but standards have sometimes lagged the development of new industries and sometimes they haven't taken hold at all. Who knows where we would be if the world had settled on a global standard—either 50 hertz or 60 hertz—years ago? Still, UL standards managed to make plug and play work for electric appliances inside homes and businesses, if only continent by continent. Groups like NIST, SGIP, ANSI, IEC, and IEEE, on the other hand, seek to drive *global* interoperability standards for the smart grid that will one day make plug and play ubiquitous on the industrial side of the grid.

However, even as standardization has earned its place in nearly all industries, providing a stimulus for mass adoption and lower costs, the debate continues as proponents of a proprietary approach promote benefits such as innovation and quality control. By building exclusivity into product design and business processes, the proprietary approach creates a buffer between the company and the outside world that allows not just control, but a revenue stream to finance innovation while maintaining quality.

Preventing interoperability is a fair trade-off in this worldview, if innovation and quality control result in superior products and more customer value to drive market adoption. Steve Jobs chose to integrate his Apple software and hardware innovations into a complete user experience with the revolutionary Macintosh, accepting niche status in exchange for what he saw as a superior customer user experience. And he's hewed to that path ever since, most recently with iTunes, the iPod, the iPhone, and the iPad. While it's hard to argue with Job's version of proprietary success, the success story of standardization seen in the IBM PC clone shows how an industry standard hardware design and common Microsoft operating system software forged a future that guaranteed ubiquitous computing. And in the digital entertainment and communication sector, a range of alternative products hew to various standards in their existence outside of the proprietary "iEconomy." So, this dichotomy, or this dual path between standardization on the one hand and proprietary approaches on the other, coexists in the marketplace.

To date, proprietary devices have certainly been more the norm rather than the exception on the industrial side of electric utility operations. The opportunity unfolding that is driving the advanced smart grid is the growing realization that an adherence to proprietary products and processes has become a stumbling block to electric utilities, and while proprietary approaches are likely to persist for a long time to come, the electric industry has entered the age of standardization when it comes to the smart grid. Advanced smart grid projects can be expected to take advantage of that momentum.

Integration

Electric utilities are organized traditionally in silos based on their functional areas (i.e., generation, transmission, distribution, metering, and retail services) and while the silos work together to ensure reliable operations in the electricity supply chain, this structure begins to work against the utility when it comes to the integration of applications and operations over a network. In fact, the standard practice of procuring systems that communicate using different physical networks (many proprietary and non-IP), that store data in separate databases, and that require separate support systems has become a principal concern and obstacle to implementing a smart grid. Such functional separation, which once provided the benefit of focus and enabled specific functional needs to be addressed, now works against a utility that seeks to integrate operations and leverage a common network and common database to provide improved interoperability while reducing costs.

With more and more granular decision making, utility managers need to manage complex and growing databases, and they need access to a universal set of timely data, as well as visibility of system operations of the entire organization. Data can flow to and from a common database to enable individual applications to draw on a comprehensive set of timely data. New operational efficiencies can be identified. An integrated network ecosystem, in contrast to silos, blends the activities inside the utility and among vendors that serve the utility to promote synergy between utility operations, utility communications, application vendors, and network vendors. When viewed from a more holistic perspective, as an integrated ecosystem rather than a collection of partially connected silo organizations, the utility gains tremendous efficiencies. To achieve these and other goals, a focus on the IP network and an accompanying network management system is needed. To create a new energy enterprise architecture and operating system and enable seamless deployment and management of a variety of applications, utilities must begin with a managed network.

Analog-to-Digital Transition

A key driver of the changes under discussion herein is a transition now underway throughout the utility industry, in which analog devices and processes are swapped out for digital versions that can do the same thing, only better and cheaper. Human beings using analog equipment to gather information and make business decisions are being replaced by automated processes using digital equipment and digital communication systems. One industry expert described the changes thus: "Utility control rooms," he said, "once staffed by 50- or 60-year-old industry veterans with 30 years of experience who personally knew all the secrets of the utility—where all the skeletons are buried—those control rooms will in 10 years be staffed by 26-year-olds with graduate degrees, who may have only paltry field experience and little knowledge about the utility service territory, but who will know well how to operate digital distribution management systems, network management systems, and the like, and who will be far more comfortable in front of a computer screen than out in the field, up a pole."

A good part of this transition away from analog instruments and people, in no way unique to the electric utility industry by the way, is based on the maturing Internet, advances in communication systems, and increasing value from Moore's law and Metcalfe's law, which we'll mention again and again throughout this book. Together, these forces drive digitization and its related efficiencies and improved customer value propositions that all work to embed digital technology ever more deeply into our lives. Changes occur not just at the network level, but perhaps more significantly with the upgrading of end devices and sensors and the adaptation of business processes to these new capabilities, architectures and operational designs.

Consider, for instance, the remote terminal unit (RTU) [6]. The RTU is essential to the operation of a utility substation and the *thousands* of dollars for each RTU have long been considered worth the investment. But on the horizon are digital routers that will replace RTUs costing only *hundreds* instead of thousands of dollars, and have greater functionality and versatility to boot. Such change will not just revolutionize the cost structure inside utility distribution networks—as digital transitions like this are implemented, they will free up enormous amounts of capital that then becomes available to finance other digital transitions, creating a virtuous cycle of change. The change, once it begins, will become viral.

But as other distribution industries (e.g., the recording industry, the publishing industry) have learned to their chagrin over the past 15 years, technology-led efficiency can be a blunt instrument that brings with it significant disruption. Such dramatic change must be managed and planned for, so that the medicine meant to cure the patient does not instead kill the patient. What lessons are to be learned from previous significant digital transitions?

One lesson is that a niche of specialized applications will keep the old technology around. The DVD resulted in more perfect reproduction of sound, but audiophiles have held on to their record collections, insisting that DVDs lack the tonal qualities of music recorded and played on vinyl LPs. The Kindle [7], iPad [8], and other digital readers have spurred a shift to digital books—Amazon announced in mid-2010 that it now sells more digital than hard copy books—but few expect the book to go away anytime soon, and book lovers are not likely to give up the textual sensation of turning a page of a beautifully bound book in exchange for being able to have an entire bookshelf in a 2-pound device. Analog devices and human processes will remain for a long time in utilities, no doubt, as those processes that have sustainable value propositions retain adherents. However, the electric utility industry is on the cusp of a tremendous restructuring as each utility reevaluates its core business functionality and asks itself what the nature of this digital transition should be—to what extent should old equipment and processes be replaced by new ones? The move to implement advanced smart grids will demand this evaluation be completed; analog devices will be replaced by digital devices only after thorough research and business process improvement (BPI) projects that follow, so that utilities may adapt operations to these new realities.

Two Axes: Functional Systems and Network Architecture

To understand the changes technology brings and how the grid will transform, it is helpful to look at two axes that create a change matrix (Figure 2.1). First, the utility business and the emerging smart grid can be segmented along the lines of utility system functions, ranging from central generation and the distributed control systems (DCSs) that enable efficient generation dispatch to the other end of the spectrum, where emerging DER systems (solar PV systems, electric vehicle charging systems, and so forth) and the smart inverters that will connect them to the smart grid portend a revolution in electricity delivery and consumption. Second, the smart grid system components run from one or more supporting networks on one end to back office servers and databases on the other. The advanced smart grid matrix (Figure 2.1) describes 96 different cost components—four sets of 24. The following section describes the advanced smart grid system that must emerge over the coming decade to support utility operations along these systems and functional areas and with these smart grid system components.

Systems and Functional Areas

Distributed Control System (DCS)

DCS is used to connect the central power plants of a utility with its control center for generation dispatch. This component of a smart grid project involves provisioning high-speed connectivity, generally fiber optics, between the plants and the energy control center and applications that enable interoperability and automated response using separate or shared databases.

Advanced Smart Grid Matrix	SYSTEMS AND FUNCTIONAL AREAS							
	Generation	Transmission/ Distribution		Revenue/Billing		Distanced Energy Resources		
	DCS	EMS/ SCADA	DA/SA	AMI	DR	DG	EV	ES
NETWORK								
Spectrum	1	2	3	4	5	6	7	8
Network Equipment	9	10	11	12	13	14	15	16
Backhaul	17	18	19	20	21	22	23	24
END DEVICE								
Hardware	1	2	3	4	5	6	7	8
Software	9	10	11	12	13	14	15	16
Network Operating Software	17	18	19	20	21	22	23	24
BACK OFFICE								
Hardware	1	2	3	4	5	6	7	8
Software	9	10	11	12	13	14	15	16
Network Operating Software	17	18	19	20	21	22	23	24
ANCILLARY SERVICES								
Project Management	1	2	3	4	5	6	7	8
System Integration	9	10	11	12	13	14	15	16
Training	17	18	19	20	21	22	23	24

(Left vertical label: SMART GRID SYSTEM COMPONENTS)

Figure 2.1   Advanced smart grid matrix.

Energy Management Systems and Supervisory Control and Data Acquisition (EMS/SCADA)

EMS/SCADA systems are used to bring back data from distributed elements of the transmission and distribution system for monitoring and control. Subcomponents include remote terminal units (RTUs) and programmable logic controllers (PLCs). This component of a smart grid project involves provisioning high-speed connectivity, generally fiber optics, between the end points on the network (the RTUs, PLCs, and so forth) and the energy control center and applications that enable interoperability and automated response using dedicated or shared databases.

Distribution Automation (DA)

DA includes three principal subsystems: a *geospatial information system* (GIS) integrated with an *outage management system* (OMS), which will ultimately be replaced in smart grids with a *distribution management system* (DMS). These systems work together to automate distribution system monitoring and control. This component of a smart grid project involves provisioning high-speed connectivity, generally wireless communications, fiber, or broadband over power line (BPL) communications, between the end points on the network—fixed and mobile utility assets—and the energy control center and applications that enable interoperability and automated response using dedicated or shared databases.

Advanced Meter Infrastructure (AMI)

AMI is comprised of smart meter end devices, a wireless communication network, and data backhaul network, integrated to provide interval consumption data collection and processing for use in revenue metering and bill production. AMI systems also provide ancillary functionality, including outage management information and remote turn on/turn off. This component of a smart grid project involves the deployed smart meters, network connectivity (generally wireless, PLC, or BPL communications) between the end points on the network and the energy control center, and any hardware and applications that enable interoperability and automated response using dedicated or shared databases.

Demand Response (DR)

DR systems consist of a remote control unit connected to a wireless network, used to automate load curtailment as an alternative to dispatching additional supply resources. This component of a smart grid project involves provisioning telecommunications connectivity, generally wireless or BPL, between the DR device, generally a home energy management system (HEMS), smart thermostat or a load controller, and the energy control center and hardware and applications that enable interoperability and automated response using separate or shared databases.

Distributed Energy Resources (DERs)

DERs are premise-based systems that produce, store, and/or manage power at the edges of the grid. The principal DER categories include distributed generation, electric vehicles, and energy storage systems. Each of these DER elements includes some combination of metering and submetering, customer portals, in-home displays (IHDs), building energy management systems (BEMSs), and home energy management systems (HEMSs), to support functionality at the ends of distribution feeders. This component of a smart grid project involves provisioning high speed connectivity, generally wired, or wireless, or BPL communications, between each single element or a digital premise server (DPS) and the energy control center and any hardware or applications used to enable interoperability and automated response using dedicated or shared databases.

Distributed Generation (DG)

DG includes any variety of edge-based electricity producing technologies and devices, far more in number, but with far less capacity per unit than traditional power plants. The most popular example of DG today is rooftop solar photovoltaic (PV) systems, consisting of solar PV panels, inverters, and net meters.

Electric Vehicle (EV)

EV includes the electric or hybrid electric vehicles, electric charging stations, and respective supporting networks. Charging stations are likely to be deployed at residences and businesses, as well as at public locations including charging stations available to the public at curbside, parking garages, and parking lots.

Energy Storage (ES)

Energy storage is becoming available at the premise, community, and utility-scale levels, but the principal distributed energy storage in the near term is likely to be *community energy storage* (CES), which is midrange in size and capability and serves multiple collocated residences or businesses.

Smart Grid System Components

Spectrum and Network Equipment

This category starts with the need to lease licensed *spectrum* (if unlicensed spectrum is used, there is no need to lease licensed spectrum). The network equipment functional area depends on whether the utility pursues a strategy of dedicated assets and builds its own network or a strategy based on outsourcing and buys network services, sharing carrier network assets. Network functions are comprised of two parts: (1) *last mile network equipment* such as wireless or BPL (this category is the full complement of installed network equipment), and (2) *backhaul system,* which provides for the necessary data backhaul between the edge network and the utility core.

End Device

The end-device functional area is comprised of three parts: (1) *hardware* such as smart meters, which includes the hardware used for specific applications at the edge of the network, (2) *application software,* which includes the application software needed to achieve end device functionality, and (3) *network operating software,* which includes the device software needed to manage the device and connect to the utility network operations center (NOC).

Back Office

The back office functional area is comprised of three parts: (1) *hardware,* which includes the cost of hardware such as servers and data storage devices, (2) *software,* which includes the utility application software such as SCADA/EMS, databases used to collect and manage data from the head end of the network, and system integration middleware, and (3) *network management system,* which includes the software used to manage the network from the utility NOC.

Ancillary Services

The ancillary services functional area is comprised of three parts: (1) *project management,* which includes project management to implement a smart grid project (including software, consultant contracts, labor, and overhead (2) *system integration,* which includes the process to make the different systems interact (including consultant contracts, labor, and overhead), and (3) *training,* which includes training made necessary by the implementation of a new system.

The New Rule of Integration

A new rule of integration is needed: *As we move from application silos to integrated ecosystems, applications procurement by utility divisions must meet network realities.*

Spend any time at all around a vertically-integrated electric utility and it rapidly becomes apparent that the traditional organizational culture is distinctly defined by four major areas of operations: generation, transmission, distribution (often shortened to just G, T & D), and retail operations (i.e., metering, billing, and customer service). These separate functions combine to create the electricity supply chain, and in the slowly changing environment of the past 100 years, this system has worked out quite well for the electric industry and the markets it serves. In certain regions though, utilities have seen their operations *unbundled* to facilitate transitions to competitive markets, and the generation and retail operations functional areas have been separated to create altogether unique companies.

When all goes well, the vertically-integrated utility is like an orchestra whose parts work together seamlessly to produce beautiful, harmonic music. However, as new functionality is added to utilities (i.e., DER), new areas of operations will raise an important question for utility managers. Should these new and growing areas of operations become new silos that increase complexity further and require significant integration efforts? Or should the addition of such new and different functions challenge the organizational structure of utilities, leading utility managers to consider new options and business models?

In fact, faced with a more dynamic environment over the past several years, and more important, the potential to leverage network technologies and new architectures by working more closely together, this distinct functional separation has become more of an impediment than a boon. We've discussed the issue with silos previously, but now we must dive deeper to understand the true impact of transitioning from vertical silo operations to horizontal integrated operations. The word "silo" may be loosely defined as "a unique area of contained operations that lacks cooperation and coordination with other areas within an organization." While the silos have interlinked operational processes, certainly, designed to enable them to accomplish their organizational missions, they often operate independently when it comes to administrative matters, financial measurement (P&Ls), planning, purchasing, and other areas. A common complaint is the duplication of effort that comes from a lack of coordination and poor communication between silos. From a technical perspective, the principal concern is the implementation of systems that require separate support systems, from communication networks to back office IT systems and databases, as described in Figure 2.2.

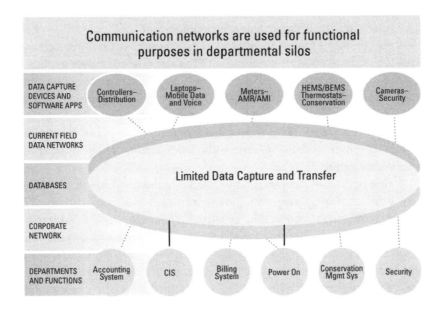

Figure 2.2 Departmental silos and support systems.

A significant consequence of organizing in silos is seen in the utility control center, where multiple screens running multiple applications provide controllers separate, distinct views of utility operations, often in different formats, which they must then integrate manually to get a more comprehensive view of the utility system's operations. Unlike telecom operators today, electric utility operators have no view of their entire operations on a single screen. Unlike the telecom operator who sees when a cell phone enters or leaves the network, an electric utility operator has no ability to see single devices; in fact, the operator is blind to grid events beyond the distribution substation. Remarkably, electric utilities still rely upon individual customers to phone in to notify them of an outage in order to determine the extent of a utility outage in any detail out at the edges of the distribution grid.

To examine the silo issue in greater detail, let us consider the application procurement process in a utility department. It is still quite typical for an application to be purchased based on specific departmental (silo) requirements, with little to no coordination with other departments or with the utility IT or telecommunications department. Savvy application vendors bundle their products into packaged solutions, complete with end-device hardware and software, a proprietary communication network, and server and database. When multiple departments in a utility follow this process, over time multiple communication networks accumulate; multiple back office systems proliferate; support costs climb as all these networks must be maintained, spare parts must be purchased, and so forth; and finally complexity mounts as schemes to achieve interoperability must be designed and redesigned over and over.

With the advent of new network architectures, operations, processes, equipment, and software, a new alternative has emerged to replace application silos—the integrated network ecosystem. Figure 2.3 describes the potential for a new dynamic process that blends the activities inside the utility and among vendors that serve the utility, highlighting the synergy potential of business process improvement (BPI) using four quadrants: on the top, utility operations and utility communications support; and on the bottom, application vendors and network vendors.

With a new set of interoperability standards and processes that stress efficiencies from a holistic perspective (looking at the organization as an integrated ecosystem rather than a collection of partially connected silo organizations), tremendous efficiencies become available. Data can flow into a common database, which enables individual applications to draw on a comprehensive set of timely data rather than more limited subsets that risk leaving blind spots to grid managers. Operational efficiencies can be identified when a complete picture of grid operations becomes available. Consider that a mere 3% improvement on a $1 billion dollar annual operational budget produces a $30 million dollar

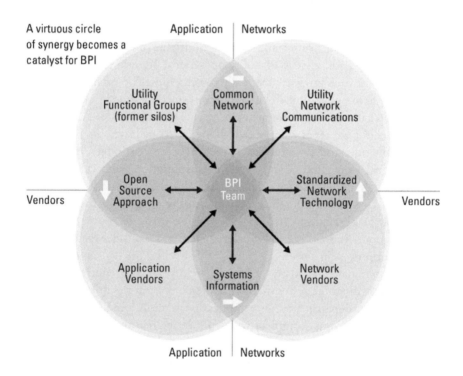

Figure 2.3 Synergistic business process improvements.

reduction in operational expenses—each year—that can be used to defer a rate case, finance other operational efficiencies, enhance shareholder returns, or retire debt.

Integration of Utility Communications Networks and Intelligent Edge Devices

The current communication ecosystem in a utility is characterized by several disaggregated utility networks operated independently in support of specific applications. Figure 2.4 details the separate systems that comprise a first generation smart grid, in support of multiple applications organized in traditional silos. Moving from left to right, note that each system brings with it a separate network to connect the field applications and data with the back office of the utility. Each of these six separate networks support specific applications within the silos that comprise the different functional areas of a vertically-integrated utility, but to function within the utility system, they must be integrated with each other, which requires special software and multiple integration projects.

In this scenario, a system-wide fiber network deployment supports both the DCS that manages the utility power plants and extends out to the substation

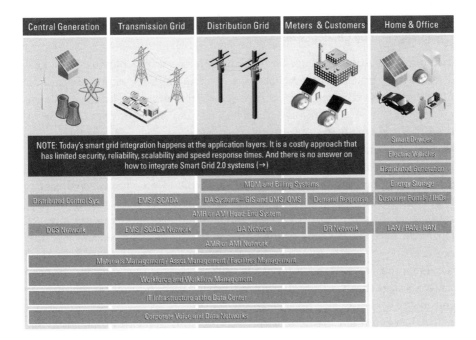

Figure 2.4 First generation smart grid.

level to support the EMS/SCADA system and transmission interconnections. DA combines multiple systems, some of which are linked: the GIS is a static asset management system that currently lacks dynamic input into the smart grid; the OMS, a reactive work management system to manage an outage crisis, draws asset data from the GIS and correlates incoming phone calls reporting outage events to work orders for truck rolls; and the DMS, a new system that will ultimately come to replace the OMS, is comprised of sensors on the distribution gird that communicate through a narrowband wireless network (e.g., 400 or 900 MHz). Likewise, the utility's two-way AMI system is supported by a 900-MHz dedicated wireless network. The DR system uses a narrowband network that interconnects tens of thousands of smart thermostats with the utility.

While few if any utilities have deployed these next systems at any scale yet, the new family of DER technologies depicted in Figure 2.4 are coming online in the next few years and must be integrated as well. This group includes DG (principally solar PV rooftop systems), EVs (including charging systems) and energy Storage. Each of these DER systems employs an inverter with a proprietary communication module that will need to communicate with the utility, likely through a shared narrowband wireless network under current operating procedures. This sample first generation smart grid deployment, using this step-by-step incremental approach, is estimated to take 5 years or more to build and to require more than 200 separate integration projects.

Power Engineering Concept Brief

Two principal challenges have become apparent so far in deploying a first generation smart grid. First, the challenges of maintaining a secure network are made even more challenging by this incremental approach. Second, incremental deployment of application-led systems requires numerous system integrations projects, which are complex and costly.

The distinct networks in a first generation smart grid have unique and proprietary security, different service level agreements (SLAs), different speeds, different coverage and different costs. In short, they are a complex challenge to deploy, much less to manage. Utilities are purchasing their solutions by selecting applications first, optimizing on the specific functional solution without much regard for network efficiencies and application integration costs and probably not realizing the duplication and complexity they are creating, the limitations they are imposing, and the unnecessary risks they are taking by having so many networks to manage, with no device interoperability and more important, no end-to-end security.

The operating expense of extra networks, not to mention the additional human resources required to manage and support the networks, is taxing and wasteful, and grows more so as more devices are added. And given the wide variety of technologies, there are no standard network or performance management tools for most of the networks. Furthermore, most solutions available today do not offer end-to-end cyber security—meaning device security plus software security plus network security plus utility NOC security—all of which are required and need to be integrated.

In summary, the current proprietary networking solutions of a first generation smart grid are incapable of providing essential qualities needed going forward; full quality of service (QoS), virtual private network (VPN), intrusion detection, and firewall capabilities. Utilities need a true end-to-end smart grid solution for substation automation, distribution automation, distributed generation management, load control, demand response, and advanced metering, which must include the following benefits: (1) the strongest security protocols and standards available, including PKMv2, CCM-mode AES key-wrap with 128-bit key, EAP/TLS (with x.509 certificates), and IKE/IPSec, and (2) intelligent, standards-based remote smart grid device monitoring and management via a proven, adaptive, smart grid network management platform and network operating system.

System integration is a huge challenge. To build a smart grid under first generation smart grid conditions, one must engage in integration at the application layer, which requires the purchase of a middleware solution and interconnection of all the corporate and engineering applications to be able to provide management oversight. With each new application added, the number of system integration projects grows according to the equation $(N^*(N-1))/2$, where N = number of systems. As N grows larger, the number of projects becomes costly and unmanageable, and the security of the system is challenged. An alternative to this approach is to implement an enterprise service bus, a considerable expense, but worth it to avoid the growing system integration expense.

Another costly project element is oversight, which requires the ability to capture key performance indicators (KPIs) and to provide reporting and decision-making dashboards for operational managers and executives in the company. Such integration projects are not trivial—their cost is about twice the cost of the software and the hardware that runs the software. And labor commitment remains considerable, even after paying high-priced consultants to complete the project, as staff must be trained to take over, maintain, and run the solutions. Worse yet, so much system integration weakens the security framework as well. As discussed in detail above, integrated device security is difficult to achieve, as most devices require proprietary retrofitting after the deployment of each of the original networks, making NIST and NERC CIP compliance very difficult, if not impossible.

In contrast to the first generation smart grid approach, consider Figure 2.5, which depicts the entire integrated ecosystem of a second generation smart grid. This approach differs from the previous assessment by using an integrated

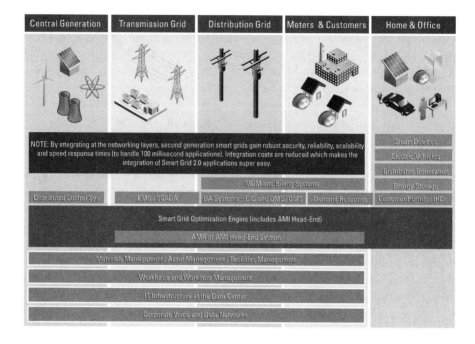

Figure 2.5   Second generation smart grid.

network approach as depicted in dark gray, which describes a smart grid optimization engine (SGOE) used to manage the advanced smart grid [9].

The second generation approach begins with a smart grid architecture design, with focus on the IP network(s) and an accompanying network management system, to enable seamless deployment and management of a variety of applications.

The Advanced Smart Grid Approach

A supporting IP network(s) and smart grid optimization engine are fundamental to the success of an advanced smart grid project. Wired IP technologies available include fiber, BPL, and Ethernet; wireless IP network technologies include 3G and WiMAX, available today, and LTE, predicted to be widely available as soon as late 2011 or early 2012. Regardless of the technology choice, IP network infrastructure is used to support all the systems in Figure 2.5, including DCS, EMS SCADA, DA (includes substation automation), AMI, DR, and DER (i.e., DG, EV, and ES).

Power Engineering Concept Brief

The process involved in deploying an advanced smart grid is noteworthy in three key areas. First, the utility pursues access to an IP network capable of supporting all of its communication and application needs going forward. In this build versus buy decision, the utility has three options: build and own a network or networks, lease space on a commercial network, or the third choice and probably the most likely: choose a hybrid of the two.

Second, the utility leverages a smart grid optimization engine that enables it to avoid the multiple integration projects required in the first generation approach above while building the advanced smart grid, but also to do much more. The smart grid optimization engine provides dynamic balancing of volt/ VAR levels based on real-time data inputs from a multitude of devices. But the smart grid optimization engine also provides the ability to control the devices and the grid in real time. The smart grid optimization engine anticipates a much more complex environment, where two-way power flow occurs as the norm rather than the exception.

Finally, the utility leverages standards-based digital devices in the field, substituting for proprietary devices that it had previously relied on. For example, RTUs costing thousands of dollars will be replaced by digital routers costing hundreds of dollars. The DCS and EMS SCADA systems in the second generation approach will be connected by fiber networks, but the remaining applications will all be supported by a more efficient integrated hybrid broadband network solution.

A New Energy Enterprise Architecture and Smart Grid Optimization Engine

Why are a new architecture and smart grid optimization engine needed for the advanced smart grid? In order to manage complex and growing databases, more and more granular decision making is required to use that data, so utility managers need an architecture that provides access to a universal set of timely data, and visibility of system operations of the entire organization. Accuracy and timeliness depend not just on which database the data is drawn from, but when that database was refreshed and so on. Without such a management system, utility management has what one utility executive has described as "ten thousand versions of the truth." At any particular point in time, a utility manager in an energy control center must ask, "What is real, right now?" With inadequate, incomplete, and/or out-of-date information, the definition of reality becomes skewed and highly subjective. At a minimum, management decisions that rely on a subjective interpretation of reality lose effectiveness, with risks escalating from there.

If the vision is to achieve the virtuous circle of business process improvement derived from a holistic integrated energy ecosystem as described in Figure 2.3, then the current systems of disaggregated utility network communications and disaggregated databases inside a utility in Figures 2.2 and 2.4 must be brought under a smart grid optimization engine with an integrated, common database schema. While noting that multiple networks may persist, it is critical that the network traffic be managed from a central point. Only with a smart grid optimization engine as described above will managers working in an energy control center achieve unitary management vision and control.

Power Engineering Concept Brief

Utilities today have a fragmented view of operations derived from the silos approach and dependence on proprietary technologies that lack the ability to communicate with each other. Beyond operations, the fragmented view impacts utility system planning as well. At the beginning of each week, electric utility managers design on paper an electric network model based on anticipated conditions, which describes the current status of all the systems that comprise the distribution grid, but the planned design they envision is not maintained throughout the week. In fact, walking through an energy control center today would show multiple operational units monitoring and managing different parts of the grid, from DCS to EMS/SCADA to OMS to AMI to DR, each with a distinct view of the state of the grid provided by the stand-alone proprietary systems. It is left to the human grid operators in the control center to integrate these disparate views of the grid and make management decisions with the information they have at hand.

The electric network model for any utility is the logical representation of the interconnection of electric elements—resistors, inductors, capacitors, transmission and distribution lines, transformers, voltage sources, current sources, demand devices, and so forth. This model interprets the interaction between the elements of the network based on the rules of physics, including Kirchoff's law [10], Ohm's law [11], Norton's theorem [12], and Thevenin's theorem [13]. The system model anticipates voltage, current, and resistance to help grid operators anticipate system impacts. This method of system planning depends upon a defined service territory with a limited number of devices—it monitors the wires and connected devices out to and including the distribution substation. In short, this method relies on a relatively simple model of the complex grid system, projects a static plan, and then changes it based on events during the week. This system of relatively manual operations planning can persist because the number of devices under management still remains relatively limited. Manual system planning is deemed "good enough" now, but it will quickly become inadequate with the increasing number of edge devices that make two-way load and energy flow a reality.

Features and Benefits of an Integrated Energy Ecosystem

A smart grid optimization engine, as described above, becomes an essential component of the advanced smart grid. Let us explore in more detail the features of this new visionary tool.

First, a smart grid optimization engine would need to provide *universal management* functionality; it should be capable of running on any conceivable IP network (i.e., wired networks such as fiber and Ethernet, or wireless networks such as 3G, WiMAX, or LTE).

Second, it is critical that the smart grid optimization engine provide *complete security* that is NIST, NERC CIP, and FIPS compliant; it should support end-to-end security, from the devices at one end, on through the software running on the devices, to the network transporting the data, down to the utility NOC presenting the data.

Third, the smart grid optimization engine should be capable of operating at near real-time speeds—at 100 milliseconds or less—and be able to fully support Internet Protocol (IP). *Instant communication* will be needed to support the functionality of an advanced smart grid.

Fourth, a smart grid optimization engine should provide *superb interoperability*; it should be able to support all electric devices (e.g., transformers, feeders, switches, capacitor banks, meters, inverters) from any vendor, because utilities are unlikely to settle for a reduced set of options when it comes to finding the right devices and applications to run their grids.

Fifth, a smart grid optimization engine should be capable of growing to meet future needs. Such *massive scalability* will be needed—when the distributed energy resources now under development become commercially viable and begin deployment, millions of new devices will come under utility management purview.

Finally, the smart grid optimization engine deployed to run the advanced smart grid must not only be affordable, it has to be economically competitive on a total cost of ownership basis: it must be more affordable than the dedicated multinetwork solution it intends to replace and offer the *lowest total cost of ownership* (TCO).

Beyond features, what *benefits* would be expected to derive from such a smart grid optimization engine?

First, the smart grid optimization engine would be expected to provide *enhanced energy efficiency*, not only improving distribution grid reliability and power quality but also reducing distribution line losses.

Second, the smart grid optimization engine would certainly provide *improved operational efficiency*, based on new capabilities such as real-time monitoring and control at the NOC level, self-healing network functionality on the grid, and adaptive distribution feeders managing the distribution circuits all over the utility service territory.

Third, the smart grid optimization engine would offer *greater customer satisfaction*, as proactive outage and restoration services were enacted, as enhanced energy products and services were made available, and as retail energy products were bundled based on targeted customer needs.

Fourth, the smart grid optimization engine would contribute mightily to societal and utility goals for a *gentler utility environmental impact*, whether from reduced or sequestered CO_2 emissions, better use of existing infrastructure, closed fossil fuel plants, or from leadership in meeting regulatory requirements.

Finally, the smart grid optimization engine would provide tremendous economic benefit, as it *reduced capital and operating budgets* based on its more effective use of system inputs and infrastructure.

A Future of Robust Digital Devices and Networks

A short way to describe the necessary transition to an advanced smart grid could be as follows: *When everything becomes smart and networked, the traditional utility becomes the advanced smart grid.*

When edge devices become "smart," equipped with IP network communication functionality and localized intelligence, they become capable of being programmed to operate independently and managed remotely, and bringing back levels of detailed information about the status of the grid never before seen by utility managers and energy consumers. In short, intelligent edge devices suggest a dramatic transformation in grid management capability, as processes designed for maintaining grid stability in the absence of information must be replaced by processes designed to leverage an abundance of information.

In other words, the current grid is designed to operate with a lack of intelligence, using assumptions and estimates that accommodate ignorance and rely heavily on human intuition and intervention, as well as a blend of proactive processes such as tree trimming combined with reactive management processes such as outage restoration. The advent of connected intelligence portends new approaches that accommodate *intelligence* rather than *ignorance*, rely on *digital connectivity* and *automation* rather than *human intuition* and *intervention*, and *proactively diagnose and repair problems* to a far greater degree, rather than *reactively responding to crises*. The grid will never be able to eliminate reactive management altogether, given the impact the environment has on the distributed grid. But an advanced smart grid promises to dramatically reduce reliance on reactive processes in favor of far more proactive behavior.

As smart devices are deployed gradually throughout the grid and brought together on a IP network(s), the advanced smart grid becomes the logical and inevitable business model and architecture to replace the traditional utility business model of interconnected, relatively independent departments operating in silos.

Throughout this chapter, we've painted a picture of technology advancing on all fronts into the utility domain. The combination of smart edge devices, a ubiquitous, integrated IP network to connect them all and a network management system to enable a new set of rules, policies, and procedures describes not just the new advanced smart grid, but also smart networks emerging in other infrastructures that will converge with the grid.

In Chapter 3, we will explore the different infrastructures that support our modern lifestyle and economy and show how digitization and the drive to network intelligent devices naturally lead to *convergence*, bringing separate infrastructures together based on common processes, supporting networks, and databases. What indeed is the advanced smart grid, if not the convergence of the electric grid with the telecom network and the Internet? Advanced meter infrastructure (AMI) reforms the revenue collection mechanism of the electric utilities to more closely resemble the relatively mature ATM networks that have provided distributed banking services for decades.

Water and gas distribution systems also benefit from AMI like electric grids do; indeed, these systems save costs when they share a communications network and related back office systems with an AMI system designed for an electricity grid. Does not the advanced smart grid ensure the changes that are needed for the electricity infrastructure to support the introduction of electric vehicles into a traditionally gasoline-fueled transportation infrastructure? And advanced smart grids will lead to the convergence of the electricity grid with our infrastructure of buildings and homes—the *built infrastructure*—to incorporate both demand response through connected energy management systems and energy efficiency mechanisms.

Endnotes

[1]   NISTIR 7628, *Guidelines for Smart Grid Cyber Security: Vol. 1, Smart Grid Cyber Security Strategy, Architecture, and High-Level Requirements,* The Smart Grid Interoperability Panel–Cyber Security Working Group, August 2010, http://csrc.nist.gov/publications/nistir/ir7628/nistir-7628_vol1.pdf.

[2]   http://www.nist.gov/smartgrid/priority-actions.cfm.

[3]   http://csrc.nist.gov/publications/nistir/ir7628/nistir-7628_vol3.pdf.

[4]   http://www.govinfosecurity.com/articles.php?art_id=3260.

[5]   http://www.oe.energy.gov/DocumentsandMedia/02-1-11_OE_Press_Release_Risk_ Management.pdf.

[6]   http://en.wikipedia.org/wiki/Remote_Terminal_Unit.

[7]   http://www.amazon.com/kindle-store-ebooks-newspapers-blogs/b?ie=UTF8&node= 133141011.

[8]   http://www.apple.com/ipad/.

[9]   The SGOE is described in greater detail in this chapter and in Chapter 7. For now, it is sufficient to consider the SGOE as an advanced network management system for smart grid modeling and operations.

[10]   http://en.wikipedia.org/wiki/Kirchhoff's_circuit_laws.

[11]   http://en.wikipedia.org/wiki/Ohm's_law.

[12]   http://en.wikipedia.org/wiki/Norton's_theorem.

[13]   http://en.wikipedia.org/wiki/Th%C3%A9venin's_theorem.

3

Smart Convergence

In Chapter 2, we explored the *rationale* for an advanced smart grid, describing the transformative changes that lead new energy architecture to emerge to meet the needs of a new energy economy. We reviewed existing business practices and processes and a new set of rules, assumptions and organizing principles that arises from the new architecture. Finally, we showed how the presence of a new architecture managed by a smart network enables an array of new smart devices, and leads to more and more integration of systems to leverage the new capabilities of the network and its interconnected edge devices. A natural outgrowth of integration within the utility is the convergence of utility functions, driven by two megatrends in the industry and the economy over the past four decades—first, the analog-to-digital transition occurring in devices and processes, and second, the networking of smart devices to drive greater and greater business value, efficiency, and functionality.

Introduction

Infrastructures and technologies are converging around common issues and pressures to become smart, based on these two ties that bind. Digital and network business models, tools and infrastructures enable electricity (electric utilities), voice and data services (fixed and mobile telecommunication networks and the Internet), entertainment media (cable networks), finance (banking networks), housing (built infrastructure), transportation (vehicles, roads, rails, and so forth), water (water utilities), and natural gas (gas utilities) to become ever smarter in their own ways, each reacting to these two principal business drivers.

The electric industry has a unique position as the support infrastructure to all the other industries and infrastructures, so the changes afoot in electricity will also begin to pull at other industries, long considered to be separate. Still other industries, rather than being drawn directly into a changing utility world, will actively seek alliances with players associated with the electric industry, as it becomes more connected and more automated. The convergence we will see between the electric industry and these other industries will revolve around the ability to connect the dots between networking technologies and principles that derive from our experience with the Internet and the common thread of replacing all analog systems with fully digital systems.

Smart Convergence: Networking Infrastructures, Stakeholders, and Markets

As the Internet has shown us and as we have learned from advances in network science over the past 20 years, the value of networks grow in remarkable ways as more links and more nodes are added. Metcalfe's law highlighted an exponential relationship between network growth and network value; before the network accomplishes any work, the very act of networking, it would seem, adds value. The challenge then becomes managing through the difficulties and challenges of networking to reach the benefits that lie on the far side of the transition. Being connected has another consequence, which is to draw the newly connected nodes in the emerging network closer together and allow far greater levels of intranetwork communication. As electric utilities contemplate networking their systems in new ways, they are drawn to look at their infrastructure and management practices in new ways.

One consequence of networking is to reconsider relationships both *within* the energy system, between utilities and their power-consuming customers, buildings, and electrical appliances, and *outside* the energy system, with parallel infrastructures such as transportation networks, and water and natural gas distribution networks. The changes contemplated in this book are acting also on these other institutions, stakeholder groups, and infrastructures—the smart grid is not the only thing getting smart. We label this movement to smart internetworked systems, stakeholders, and infrastructures *smart convergence*, and we see it happening throughout the new smart infrastructure world.

As smart devices of all kinds proliferate in businesses and homes, each with incredible computing and communications capabilities, the need to monitor and control them is driving electricity managers to borrow from the telecom industry. And as the number of smart connections to the grid escalates, a new network management scheme is needed, what we call the *advanced smart grid*, described in further detail in this chapter. The advanced smart grid makes smart convergence of all types possible; consider just a few examples next.

First, the role of energy producer and consumer is converging. As energy consumers become energy producers as well, distributed generation (principally rooftop-mounted solar PV systems) creates a need to incorporate onto the grid tens to hundreds of thousands of new power electronics, inverters and net meters, all made smart by adding new computing and communication capabilities. This blending of energy consumption and production becomes the realization of Alvin Toffler's paradigm of the *prosumer* in the electric ecosystem (where one becomes *both* a consumer and a producer). If grid managers are to take advantage of the opportunity to create new resource alternatives that can be dispatched to help ensure grid stability, they will need to find a way to connect and ultimately to control these proliferating edge energy systems.

Second, the *built infrastructure* is even now evolving to gain new capabilities. Smart buildings and their occupants are beginning to take a more active role in their consumption of energy, thanks to new efficient building design and construction practices and new digital energy management technologies under the collective heading of building energy management systems or BEMS, which enable building managers to make more efficient use of the evolving electric grid ecosystem (and for home owners, the equivalent is home energy management systems or HEMS).

Third, the *transportation infrastructure* is evolving to embrace electricity-powered vehicles on a grand scale. A smart charging infrastructure in homes and businesses and out in public spaces will be needed to fuel an increasing number of electric vehicles (EVs), much as our current network of public gas stations fuels our internal combustion vehicles.

Fourth, *energy storage* promises to be a game changer, dramatically altering the capability and business processes of the electric utility, based on the smart charge/discharge functionality of associated energy control systems, as well as the location of the energy storage units and their interconnection to the advanced smart grid.

Electricity and Telecommunications

Electric and telecom utilities grew up together, kids on the same block. In some ways, these industries have been on a path of convergence from the moment they were born, as we pointed out in Chapter 1's discussion on the electric utility use of telecom technologies. Electricity made possible the invention of telephony, and in turn, the widespread network of power plants, transmission lines, and distribution lines has from the very beginning depended on voice telephony and telegraph networks. In the early twentieth century, these two industries consumed the products and services that the other produced. Later, the electric industry became a principal beneficiary of pioneering advances in telephony and IT at Bell Labs. The telecom industry and Internet pioneers that emerged in the 1990s have built data centers where they found ready access to cheap, high-quality power. For most of their respective histories, however, each industry has regarded the other as separate and distinct. In some ways, each is in the business of distributing a commodity, voice waves and later digital bits and bytes on the one hand, and excited electrons on the other, but the businesses are sufficiently different that they have remained separate industries serving separate markets.

Yet, over the past two decades, we have seen growing convergence of these two industries. Utilities began investing in fiber optics as an improvement over their narrowband wireless or copper wire infrastructure, the better to support their SCADA systems and other connected devices. Many utility leaders recognized an opportunity in the synergies that come from owning towers, poles, copper wires, and fiber lines, and created utility-owned commercial telecom units *(UTelcos)*. A couple hundred utilities now operate UTelcos inside their utilities, engaging in wholesale telecommunications service transactions including acting as a carrier's carrier and leasing dark fiber for long-haul communications, tower and pole mounting rights, and utility rights of way. A few utilities became retail telecom providers after the Federal Telecom Act of 1996 allowed utilities into the competitive local exchange carrier (CLEC) space, but few utilities have had success at retail telephony, and most of those pioneers ultimately pulled back after a handful of retail utility telecom pioneer ventures failed.

On the wireless side, the Utilities Telecom Council (UTC) [1] has advocated for favorable radio and spectrum policy for utilities worldwide for over 60 years. Utility telecommunications departments were created to manage the variety of mission critical telecom functions, from radio operations to proprietary wireless networks to the field communications needs of the different departments within the utility, as well as the more conventional telephony needs of utility office workers. Utilities now boast considerable experience as both network operators and consumers of telecom services of all kinds, comprising the second largest market for telecommunications equipment and services. While utilities debate the relative merits today of investing in telecommunications assets and infrastructure (the "build" strategy) and outsourcing their telecom needs to commercial telecommunications companies (the "buy" strategy), a new area of convergence is emerging based on new developments in software and product development.

An area ripe for exploitation by electric utilities is to reevaluate their telecommunications architecture, which has generally been built opportunistically, in ad hoc fashion to support functional applications housed in departmental silos, as described earlier. The opportunity then is to move beyond these silos, designing a new smart grid telecommunications architecture from the ground up, in order to model the successful architecture found in telecommunications and Internet networks. In other words, electric utilities should prepare for a future of networking millions of devices under the management of a single utility NOC. In a telecom world connected by ubiquitous IP networks, both fixed and mobile, where intelligent mobile devices proliferate, the potential to reconfigure electric utility telecommunications architecture is breathtaking. Utilities should anticipate an increasingly complex array of intelligent edge devices that will need to be connected, monitored, and ultimately be controlled as elements of the electricity infrastructure. This first level of convergence is foundational and deterministic to the remainder of convergences discussed below, given that it is the basis of what we call the advanced smart grid.

Power Engineering Concept Brief

Implementing a wired and wireless IP network is the root of a telecom convergence strategic project, but in contemplating such a project, one is led to ask: "If generation capacity is built to meet peak demand, then why isn't a similar approach taken when it comes to customer services and the associated infrastructure to deliver such solutions as demand response?" In other words, rather than aligning communication infrastructures and systems to meet peak demand, utilities more often have devised an incremental, lowest-cost approach to telecom, in order to meet their universal service requirement. Approaching telecom infrastructure from a lowest-cost basis and conforming processes and infrastructure to meet minimal requirements, rather than designing telecom architecture to meet peak requirements, is problematic on several fronts.

Consider that to achieve just one application goal—to make demand response a dispatchable resource—requires implementing a sufficiently robust communications infrastructure to enable real-time management at the equivalent of peak-driven telecom architecture. But a conservative culture and economic hard times have led managers to deploy wireless infrastructure that depends on such "affordable, reliable" twentieth-century technology selections as digital paging networks. Given that electricity consumers already live in an on-demand, real-time society and expect services, even utility services, to match current technology capabilities, building a system that relies on twentieth-century functionality, even if it is built at the lowest cost, is bound to come up short in meeting customer expectations.

To address these and other deficiencies and build an advanced smart grid, such disconnects must be addressed head on. As an industry, we forget that our decisions should be rooted not only in lowest cost, but also in terms of quality of service and reliability. Again, compare the investments to support quality processes and service levels that Japanese grid managers have made over the same time period to those of their American counterparts. Thanks to their foresight and investments in infrastructure and technology, Japanese grid managers

now achieve a 3-minute annual disruption of power for their entire country—
that's a "3-minute System Average Interruption Duration Index (SAIDI)" in
utility-speak. In the United States, grid managers can only achieve a 120-min-
ute SAIDI. But don't blame the managers—their infrastructure lets them down.
It's as if U.S. utilities and regulators, instead of building the right infrastructure
to support a best practice use case, have become oriented to an approach based
on "safe" incrementalism. In effect, the system built in America lowers functionality standards by default down to the capabilities of the lowest-cost system,
instead of raising those standards up to make them world-class.

To enable an advanced smart grid, in contrast, individual utilities will
need to undergo a strategic transition away from this ad hoc, cost-justified
approach by adopting a planned, robust, sustainable future-proofed network
capability and infrastructure. We are of course now talking about an IP network with fiber to mission-critical facilities, to create a foundational supporting
infrastructure that is then complemented by a wireless IP network overlay to
provide universal coverage throughout the service territory. (For utilities with
large service territories that include less densely populated rural areas, different
technologies may need to be added to provide cost-effective coverage. For the
discussion in this book, however, we remain focused on the distribution grid
with sufficient population density to rely on a single wireless IP network coverage solution supported by fiber backbone and backhaul.)

Three alternatives present themselves to achieve wireless IP network coverage: (1) a private, utility-owned IP network, (2) a service contract with a
public carrier, and (3) a hybrid private-public network.

Private, Utility-Owned IP Network

The options today for a private utility-owned network must answer the first
question of licensed versus unlicensed spectrum. The advantages of unlicensed
spectrum solutions principally revolve around access and lower cost, but are
offset by disadvantages that include the risks of unpredictable congestion and
interference. Unlicensed spectrum solutions have enjoyed a favorable environment at the outset, based on the relative lack of traffic on the spectrum, but as
traffic grows from multiple applications, multiple tenants, and a proliferation
of devices, the network performance can be expected to decline, presenting a
challenge to overcome. For example, a baby monitor sitting on a windowsill
next to a meter on the outside wall may disable meter reception that relies on an
unlicensed network. Users of in-home Wi-Fi experience this conundrum when
the microwave oven disrupts their Internet access.

In contrast, licensed spectrum solutions offer an exclusive right of use
advantage, but come at a higher cost, and require spectrum availability from
the holders of the spectrum. Beyond spectrum, a private network solution must
consider the operations aspect. A network operator, either the utility itself or a

contracted provider, is required to operate and maintain the network. A hidden challenge and cost of a private network solution is technology obsolescence, which presents a risk as new solutions become normative, making spare-part sourcing and maintenance a strategic consideration. Another challenge is that rapid changes that are common in wireless telecom are out of synch with the long-term nature of utility investment, where expectations of useful life can exceed 15 to 20 years, in contrast to telecom life cycles of only 5 to 10 years.

Diving a little deeper, we next consider the strategic network design of a smart grid private wireless IP network, designed from the ground up. The demands of a machine-to-machine (M2M) IP data network are significantly less than the requirements of a commercial wireless IP network, which must anticipate mobility and high-bandwidth video uses, as well as the need to penetrate exterior walls to reach users on the building interior. Consequently, a network designed primarily to serve a smart grid, lacking such onerous requirements, can have larger cell sizes and still find signal strength at the edge sufficient to maintain connectivity and transmit the data packets as anticipated.

The essence of this innovative approach is to design and build a large cell "thin" smart grid network that provides signal strength at the edges sufficient to read smart meters (in Figure 3.1, accepting lower signal strength at −90 dBi enables a coverage area C that is 13 times greater than that of a commercial network design (A), using the same network equipment).

This breakthrough approach to building a WiMAX network is enabled by creative design parameters, targeted strategies, and adjusted assumptions on

Exponential Leverage:
Coverage Area Grows as Radius Increases

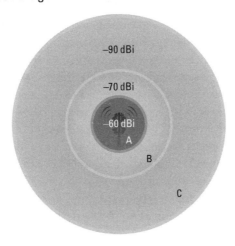

Figure 3.1   Thin smart grid network design.

such key cost drivers as tower requirements, cell radius, number of cell sites, and spectrum cost. Targeted strategies include infrastructure bartering arrangements with spectrum owners and carriers to leverage utility assets (towers, ROW, and so forth) for reduced telecom costs and the "thin" network strategy discussed above, which accepts black spots, to be covered in a subsequent commercial deployment.

When working with a commercial network operator to obtain spectrum rights, a utility can anticipate that the network equipment used in the thin smart grid network may ultimately be incorporated into a denser commercial IP network. Thus, such an initial deployment may be followed by a second phase to deploy additional network equipment to reduce cell size and raise the network capacity to meet commercial standards for providing IP network services.

Public, Carrier-Owned IP Network

The second option—a service contract with a public carrier—brings with it a new set of strategic considerations. The issues and risks of spectrum access, network operations, and technology obsolescence described above now become the purview of the public carrier, which is a big selling point hammered home by the public carriers: "Don't worry, operating, maintaining, and upgrading networks is our critical core competency." But utilities face a new challenge when considering a public carrier solution: assuming the cost of access pencils out, utilities must come to grips with the risks associated with placing a mission critical function in the hands of a third party. One utility executive explained in words to this effect: "An SLA with penalties does me no good when power remains disrupted due to a network failure. How can I explain that to a rate-payer?" So, establishing trust will be paramount if carrier options are to be embraced by utilities going forward.

Carrier options generally fall in one of two categories, depending on the network being leveraged. Carriers with maturing third generation (3G) networks, whose costs have been borne by voice and data services in the commercial space, now look at these mature networks as logical transport providers for the emerging M2M space, which have light data requirements, but which often require the kind of coverage only a carrier can provide. Since the capital costs of the 3G networks are mostly paid off, services operating on these networks can be aggressively priced based on marginal costs.

Other carriers will seek to take their emerging network technology—4G, also known as long-term evolution (LTE)—to the marketplace as a platform for both voice and data services, including M2M. The challenge in the next few years with 4G will be the immaturity of the network and the lack of coverage. Prices will be relatively high at the outset, but the coverage of the new networks will improve steadily over the next several years. The numerous technical advantages of the new 4G technology are expected to make these network options appealing and help telecom carriers to win over new converts.

Hybrid Public–Private Network

With telecom carriers aggressively marketing their mature 3G networks in the M2M space, utilities may take advantage of the coverage benefits and relative low costs of a ubiquitous 3G network to provide coverage for meter reading and other device access in hard-to-reach rural areas, and some may consider 3G as a transport solution for meter data in more dense urban areas when costs and internal capabilities make the carrier solution more attractive. However, for mission-critical functions (or urban noncritical tasks), the utility may still choose to rely on a privately owned network that provides greater certainty of performance and public reassurance of reliability, if at a greater cost. Wild cards in this area include aggressive pricing by the carriers, given the relative lack of capital for financing new network construction and the unwillingness of utility regulators to add costs of new networks to the rate base, where an existing carrier network can be used as an alternative at a reasonable cost comparison.

Electricity and IT

The second great convergence, closely related to the telecom/electricity convergence described above, relates to the steady adoption and integration of IT devices, applications, processes, and market concepts into the highly controlled and regulated world of the electric utility, which is charged with "keeping the lights on." Electric system reliability has to be one of the most challenging organizational dictates imaginable, given that high voltages are deadly, electric lines are mostly above ground and constantly subject to environmental disruption and that the highly fragmented, complex, interwoven electricity grid of 2011, built on a foundation of aging equipment, continues to grow in size, complexity, and importance as the foundational infrastructure of this digital, networked era, yet requires exacting operational standards to perform its functions.

The challenge of this smart convergence is to weed out the "good" digital adoption from the "bad." In a conservative environment, where utility infrastructure investments are subject to public scrutiny by regulatory bodies, local governing boards, and member-representative boards, replacement of working and reliable, if aging analog equipment with new digital equipment is bound to proceed slowly. Many in the utility industry hold to the "better the devil I know than the one I don't" school of thought, especially in the IT world, where IT vendors have offered an ongoing array of options that promised enticing benefits when gaining approval, but sometimes failed to deliver on those promises after they were deployed.

Moving beyond the reluctance to replace a workable solution with a new, better, or cheaper solution that may or may not work as promised, an additional challenge is a culture that has built a reliable infrastructure using industrial age business practices, where electromechanical equipment has typically enjoyed a useable life measured in decades rather than years. Beyond the relatively long useful life of electric utility assets is the business practice of leaving working assets in place when they operate well beyond the expiration of their useful lives as recorded in depreciation schedules. This culture has become a principal challenge to the digital transition in practical terms. The conventional wisdom embodied in the phrase, "If it ain't broke, don't fix it," leads to a culture of keeping assets around well beyond their documented useful life, essentially leading to a prevalent business practice of running equipment until it fails.

In contrast, the IT equipment replacement cycle is measured more in tens of months than tens of years. Driven by the economics of Moore's law, IT equipment is routinely obsoleted in a matter of 3 to 4 years based on the availability of new equipment that has greater computing power, greater storage capacity, and improved functionality, often at a lower price—in essence, making equipment *technically* or *economically* obsolete well before it becomes *functionally* obsolete. In the face of this culture gap in purchasing philosophy, IT departments must work closely with their internal clients in the utility to explore the relative merits of the purchasing cycle and a new mentality that considers opportunity costs on an equal footing with such things as spending caps, tight budgets, and traditional utility business practices.

Analog-to-digital transition drives the smart convergence of IT and electricity. An electric industry that once defined electricity as high technology at its inception has moved into a new era where the convergence of information via digital technology and electricity creates what will become a new, transformed industry over time. When this transition has reached its potential in decades to come, and electricity has finally become a blending of electricity and information, will we still think of it the same way? An industry built on estimates, best guesses, and human intuition in times of crisis is moving to a far more exacting industry that relies on facts, data, and information to drive automated processes under the oversight of human operators.

The key example of this type of smart convergence involves the replacement of older analog devices with new smart digital routers at critical points throughout the electric infrastructure, which become new intelligent nodes on a network modeled after the interconnected Web network architecture of the Internet, overlaid on the traditional radial network architecture of the electric industry. Anyone who has experienced the intermodal transportation network of a large city understands the radial network of urban railways, subways, and trolleys radiating from a city center interconnected with the web of buses and taxis traversing city streets that more closely resembles an interconnected web.

Operating together, these two networks create a synergy that is better than either could achieve on its own.

The advanced smart grid depends not only on an IP network infrastructure and sophisticated network management tools borrowed from the telecom world, but also on digital intelligence built into smart meters, smart routers, smart inverters, and smart consumer devices all acting independently on the edge based on programmable logic and algorithms, interconnected so they may be controlled remotely when needed from the central utility NOC.

The convergence of IT with electricity will lead to the following analog–digital transition. The *traditional* business practice involves human decision-making to keep the grid operational, based on deep human experience, skills, and knowledge and a healthy dose of human intuition. The *transitioned* business practice involves intelligent devices programmed with the knowledge and skills of the best practices of experienced utility workers, then executed with a precision that those human workers would never be capable of matching.

In digitally converged electric industry operations, there will be fewer workers than in the past, but the skills, knowledge, and experience of utility workers will start with traditional electricity skills of the electrical engineer, lineman, and energy control center operator and add more IT skill sets from programming to network diagnostics. New energy service jobs will arise for displaced utility workers. Business processes will need to be adjusted as well to accommodate this shift in digital capability and new labor demands of electric utility organizations. Utilities, with help from government grants and loans, will be in the business of retraining their workforces to operate grids using digital equipment.

Power Engineering Concept Brief

The installation of a ubiquitous wired and wireless network throughout the utility service territory opens the door to an examination of utility operations and the devices used to execute utility functions, for this is the tableau on which the analog—digital transition will be played out in the utility landscape. From a power engineering perspective, the digital applications with potential in the advanced smart grid paradigm include those listed in Figure 3.2, with special attention paid to the key applications described in detail next.

Condition-Based, Predictive Equipment Maintenance

As stated earlier, the prevailing business practice in most functional utility silos is to react, replacing devices at the point of failure. In contrast, predictive equipment maintenance involves searching for patterns to identify faults inside circuits. Predictive maintenance of distribution assets involves monitoring devices to detect changes in power consumption, which act as red flags. Older devices

Advanced Smart Grid Digital Applications

- Condition-Based, Predictive Equipment Maintenance/Asset Management
- Demand Response Analysis
- Direct Customer Load Control
- Dispatcher Training and Simulation
- Distribution System Real-Time Analysis Tools
- Emergency System Restoration Support
- Fault Detection
- Feeder Equipment Monitoring
- Integrated Volt VAR Control
- Integration of DMS(OMS)/AMI/GIS
- Load Forecasting
- Multilevel Feeder Configuration
- Network Switching Management/Analysis/Optimization
- Power Quality Assessments
- Relay Protection Coordination

Figure 3.2 Smart grid digital applications.

near failure consume power differently, creating distinctive patterns and signals that can be diagnosed and detected to identify problems before they happen. What's more, anticipated performance based on published specifications can be measured and compared to measured performance in real world scenarios, allowing for fine tuning and advanced diagnostics.

Demand Response Management and Analytics

Currently, DR is managed primarily during peak seasons and at peak hours in order to lower system peaks and to avoid uneconomic operations, using technology that in some cases is neither real-time nor two-way, but which still offers a valuable service to help manage the grid better during these difficult periods. Aggregation of a fleet of smart thermostats or home energy management systems, however, promises to provide the utility a powerful new resource, only now being realized. Imagine a digitally connected distributed resource of aggregated smart devices as yet another grid management tool, working collectively to fine tune and optimize grid voltage and VAR levels minute-by-minute according to preset algorithms that balance energy consumer-prescribed conditions with the utility conditions needed to harmonize the grid.

To get a bead on the potential, consider that today nearly every household has a number of charging devices, which are miniature transformers that often stay plugged in, continuing to draw power after the device has reached full charge, a condition that has come to be called "vampire power." While consumers can be urged to unplug those devices, or to buy special plug inserts to unplug them automatically, many can't be bothered with doing so. Imagine though if this waste of energy could be converted in an advanced smart grid environment into a latent asset, lying fallow in wait for a command from grid managers to switch off to help them fine-tune the grid when it is under stress.

Fault Detection, Isolation, and Restoration (FDIR)

Engineers plan the state of the grid using models, planned algorithms, and historical behavior patterns, but these are blunt instruments and the plans thus created soon go stale, since the grid actually works in real time. For instance, when a transformer located out near the end of a distribution feeder has an arcing event that puts it out of commission, the amount of load lost on the feeder line may be detected, but the cause of the loss remains imperceptible from the perspective of a control center operator, who can measure the effect, but remains ignorant of the cause. The outage will remain undiagnosed until consumers notify the utility and a truck is dispatched to locate the fault and restore the line. Similarly, when a capacitor bank cracks and becomes the electricity equivalent of a leaky pipe, it ceases to function as planned and the grid slowly grows more out of balance with low-grade degradation over time. Signal processing, the twenty-first-century cure, correlates and transmits environmental and utility functional data to allow timely comparative analysis of observed and anticipated data and an automated response to put the grid back together again.

FDIR helps achieve dramatic reductions in grid interruptions, integrates reclosers and substation equipment, and ensures better safety, but a critical challenge in restoring power—synchronization—remains. SAIFI, SAIDI, and CAIDI are measures commonly used to report electric service quality. SAIFI measures how often a customer can expect to experience an outage, SAIDI measures how long the customer can expect to wait for power to be restored (regardless of how often the system goes down), and CAIDI measures the average outage duration if there is an outage, or average restoration time. These indices are defined over a fixed time period, usually a month or a year, and can be measured over the entire electric distribution system or over smaller portions of the system, such as an operating area or individual circuit.

CAIDI is perhaps the least straightforward of the indices, but from a customer experience perspective, CAIDI is the most relevant index. While the first two indices are driven by frequency (SAIFI) or time duration (SAIDI), both variables drive CAIDI. Strategies for improving SAIFI and SAIDI can sometimes adversely affect CAIDI. SAIFI is improved by reducing the frequency of outages (e.g., tree trimming and equipment maintenance programs), and by isolating the disruption to reduce the number of customers interrupted when outages do occur (e.g., by adding reclosers and fuses). Strategies that reduce SAIFI also impact SAIDI—an avoided outage has no chance of increasing the duration number. Both SAIDI and CAIDI benefit from faster customer restoration. Perversely, system improvements can make CAIDI go up as well as down, depending on the relative impact of improvements on outage frequency (customer interruptions) and outage duration (customer minutes of interruption). Thus, it's complicated and interpreting indices requires more than a cursory glance. In the final analysis, all three indices are valuable management tools from an advanced smart grid perspective, because they represent data leveraging to enable grid managers to benchmark and make system improvements over time (Figure 3.3).

System improvement really matters today, given that modern customers have come to expect a higher level of power quality from their electric utility, as they do of the other service vendors in their lives. Digitization in part drives these perceptions as well. Higher power quality requirements have become even more critical with the proliferation of electronic devices such as flat panel TVs, DVRs, VCRs, game consoles, computers, and clock radios, which

Measuring Electric Service Quality

Three key indices measure service quality in the electric utility infrastructure:

SAIDI	SAIFI	CAIDI
(System Average Interruption Duration Index) is the calculation of how long the system remains down after outages during a specified period of time, usually a year, that is, the sum of outage Customer Minutes of Interruption (CMI) divided by the total number of customers served.	(System Average Interruption Frequency Index) is the calculation of how often the system goes down over a specified period of time, usually a year, that is, the sum of outage Customer Interruptions (CI) divided by the total number of customers served.	(Customer Average Interruption Duration Index) calculates the impact of outages on a single customer, that is, the product of dividing the duration (SAIDI) by the frequency (SAIFI).

Figure 3.3   Electric service quality.

are intolerant of even the smallest interruption of power, and can set lights to blinking throughout the house—just imagine adding solar PVs and electric vehicles into this edge power equation.

One strategy grid managers employ to limit the number of customers affected by an interruption due to a fault is to divide distribution feeders into sections then isolate them using motorized switches or breakers. A new trend is to include smart meters and distribution routers downstream in the feeder for the collection of dynamic data. In this manner, applied FDIR algorithms can detect in which section of the feeder the fault occurred and rapidly isolate that feeder section by operating the isolating switches or breakers and restoring power to the nonfaulted sections, while ensuring that only those customers on the faulted section are affected by the power outage.

Integrated Volt/VAR Control

Put simply, the purpose of the electric distribution network is to move electricity out from the transmission system to substations, then down distribution feeder lines and on to consumers. The distribution system includes medium-voltage (less than 50 kV) power lines, substation transformers, pole- or pad-mounted transformers, low-voltage distribution wiring, and electric meters. The distribution system of an electric utility is a complex system, which may have hundreds of substations and hundreds of thousands of components. Most of the energy loss occurring on the distribution system is caused by resistance: electric current flowing a distance through conductors results in a loss of measured ohms. The amount of the loss is proportional to the resistance and the square of the magnitude of the current. Thus, operators reduce losses by reducing the resistance or the current's magnitude. The resistance of a conductor is determined by the resistivity of the material used to make it, by its cross-sectional area, and by its length, none of which can be changed easily in existing distribution networks. Fortunately, reducing the current magnitude can be accomplished more readily, by eliminating unnecessary current flows in the distribution network.

When evaluating line loss, there is also the issue of active and reactive power to consider. For any conductor in a distribution network, the current flowing through it can be decomposed into two types, active and reactive. Reactive power present in the line contributes to power loss by using up a portion of the current carrying capacity of the distribution lines and equipment. Reactive power compensation devices are designed to reduce or eliminate the unproductive component of the current, thereby reducing current magnitude, and thus, energy losses.

Depending on the types and mixture of loads in the system, the voltage profile on the feeders can also affect the current distribution (and loss of power), although the loss is smaller and its impact less direct.

Traditional volt/VAR control extended along the distribution feeder out to the edges of the grid has achieved reasonable improvement in the distribution grid's operational performance. Voltage regulating devices are usually installed at the substation and on the feeders. Substation transformers can feature tap changers, devices that adjust the feeder voltage at the substation, depending on the loading condition of the feeders. Special transformers equipped with tap changers called voltage regulators are also installed at various locations on the feeders, providing fine-tuning capability for voltage at specific points on the feeders. Reactive compensation devices (i.e., capacitor banks or more informally, cap banks) are used to reduce the reactive power flows throughout the distribution network. Cap banks may be located in the substation or on the feeders, and can be fixed or switched.

Traditionally, the voltage and VAR control devices are regulated in accordance with locally available measurements of voltage or current. On a feeder with multiple voltage regulation and VAR compensation devices, each device is controlled independently regardless of the resulting consequences of actions taken by other control devices. This practice often results in control actions that may be sensible at the local level, but contribute to suboptimal effects on a broader scale. More ideally, information would be shared among all voltage and VAR control devices and control strategies would be comprehensively evaluated to make the consequences of possible actions consistent with optimized control objectives. We term this new approach to distribution system management *integrated volt/VAR control.*

The accelerated adoption of substation automation (SA) and distribution feeder automation (DFA), the recent widespread deployment of advanced metering infrastructure (AMI), the growing deployments of solar PV systems, and the advent of electric vehicles together provide both the rationale and the foundation for a smart distributed control approach. All these devices provide the necessary local intelligence at the sensor and actuator levels, based on reliable two-way communications between the field and the distribution system control center, to make distributed control possible. With enhanced sensory data feedback from sensors located further out on the edges of the grid and along the distribution feeder, grid managers are equipped with the local intelligence and two-way communication capability they need to exercise finer control over a larger area of the grid. These new sensors will become even more useful when they are used to help grid operators address the new challenge of reverse power flows coming from such new DER elements as solar PV systems and energy storage devices located along the distribution feeders.

In fact, the integrated volt/VAR control transforms new DER devices from potential threats to grid stability into new, valuable grid management resources. The advent of the advanced smart grid enables utilities to manage VARs proactively through capacitor banks further out on the distribution feeder, beyond the substation, whether at the pad-mount or pole-top transformer or at the smart meter on the ends of the line.

Integrated volt/VAR control will minimize power losses or demand without causing voltage/current violations, a term that refers to the undesirable excursion from the normal operating current and voltage range for the distribution system (e.g., current that exceeds the maximum safe limit for a given conductor type, voltage that exceeds a limit considered safe for consumers, or voltage that falls short of a limit needed for normal operation). Integrated volt/VAR control solutions would have emerged sooner, if not for the lack of computational resources near the edge, which are needed to solve complex mixed-integer nonlinear and nonconvex problems in order to evaluate the loss and demand for a single functional equation. Solving such equations is at the root of maintaining balance in such a complex system. Efficiency matters to address complexity: an algorithm that requires fewer functional evaluations to find the optimal solution will be regarded as more efficient than one that requires more functional evaluations to achieve the same objective.

Electricity and Banking: Smart Meters (AMI)

The revenue meter that hangs on the outside wall (mostly), or the one that is hidden down in the basement or in a closet (more obscurely) is the utility's cash register. Wherever the revenue meter sits, it is the last device on the end of the utility distribution network, and its primary role is to measure energy *consumption* in kilowatts (with commercial and industrial accounts, it also measures energy *demand* in kilowatts). The more sophisticated commercial and industrial meters also measure other qualities of the electricity that flows from the grid into the customer premises. Throughout its history, the utility has managed a relatively simple analog system to take monthly readings and calculate energy use based on the difference between two readings at the beginning and end of a billing period. With AMI, the issue of remote revenue management using a network converges with the use of ATMs by the banking industry, providing us with a discussion on technology convergence contrasting the mature network of ATMs with the dramatic expansion of networked revenue meters on the distribution grid.

What could ATM have to do with AMI? They both automate the delicate task of revenue collection over a distributed network. The ATM network got its start in 1973, when a company called Docutel [2] was awarded a patent for its networked ATM. The trend spread and the machines grew more sophisticated and numerous. Today these secure cash dispensing facilities dot the landscape and we think little of the revolution they represent. ATMs managed to take the functions of the bank vault and the human teller sitting behind a barred window, which had earlier moved from the bank lobby out to the drive-up window, and distribute them out to the edges of a large network, automating the teller function, but also providing bank customers secure network access to their bank accounts. As such, this transition offers a model of both digital transition and transition to a secure network that has evolved and improved over time, but certainly well ahead of the Internet and the proliferation of Internet security risks and strategies. To round out the picture of networked banking, banks have moved to an Internet model of account access and distribution of bank transaction statements in order to reduce operating expenses and provide more convenient customer access to banking information.

The emergence and growing and now widespread adoption of wireless networked revenue meters by electric utilities offers similar potential to transition electricity consumption monitoring and consumer access away from a twentieth-century model of manual reading of analog meters once a month to produce a paper bill to be mailed to customers for manual payment.

The change potential in automating revenue meter data collection with an advanced meter infrastructure (AMI) is dramatic. Imagine moving from a single monthly meter read (12 reads each year to produce a monthly bill) to a meter read every 15 minutes to produce a bill, but to do so much more: Four reads per hour—15-minute *interval reads*—produces 96 reads each day. That's eight times as many reads in a single day as was produced in a year with the manual, analog system. In a year, the number of meter reads will go from 12 to 35,040. And that's just for a single account. For a utility with one million residential meters, the number of interval meter reads each year by that single utility will be over 35 billion. If each single meter produces 400 MB of data per year, then an electric utility with 1 million residential meters will have the challenge of managing 400 terabytes of new data each year. And that's just for one midsized utility.

What will become of all that digital interval meter data? Besides being used as a resource to produce monthly consumer utility bills, the massive amount of universal, detailed consumption data opens up a whole world of management possibilities. Consumers will be able to use the interval meter data in home energy management systems (HEMS), whether it is communicated directly from the revenue meter via a technology like HomePlug or ZigBee into the home to an in-home display (IHD), or communicated via the utility over the Internet and presented on a Web site and accessed via a PC or a smart mobile phone. Utilities will be able to aggregate the data to produce valuable information on system operations and consumer behavior, from a group perspective, and learn new things about how to operate their utilities more efficiently.

But for this vision to come to pass, the deployed AMI system will need to be secure. The implementation of an AMI system requires not only significant planning, but also ample time during the deployment to test and calibrate to meet the demands of security. Regulations and data privacy standards have made the utility the steward of customer data and they are obliged to maintain the meter data with care. When it comes to security, the AMI system has a model in the ATM networks that have ensured billions of safe transactions daily.

Power Engineering Concept Brief

As stated earlier, ATM networks, which connect ubiquitous ATMs and are often the target of malicious hackers, provide a useful analogy to the network approach needed for the AMI system that is part of an advanced smart grid. ATM networks have managed to remain well-protected and reliable despite such persistent threats because of the highly sophisticated, standards-based, device-level security that resides in each ATM, which renders it inoperable upon threat detection before any connection can be made to the ATM network, thereby avoiding virus or worm proliferation. While smart device hacks are inevitable, utilities can protect their advanced smart grid from a massive network virus or worm by implementing ubiquitous security architectures. Aligning the architecture on security begins with embedding unique, standards-based hardware and software security into every network device to prevent penetration attacks (i.e., worms and viruses) from spreading throughout the advanced smart grid network. Granular, device-level security in the ATM network quickly identifies and isolates a hacked or compromised device, limiting damage. Similarly, sophisticated, device-level security incorporated in the design of embedded communications devices ensures the most robust protection.

From a technological and engineering standpoint, it should be noted that AMI is not by any means the end of the story. The road from AMR in the 1990s to AMI in the first decade of this new century has been about increasing real-time distribution management, isolated in the meter as an end device. In an advanced smart grid, AMI undergoes a further metamorphosis, taking these changes one step further to become what we might call advanced grid infrastructure (AGI), which includes much more than meter data collection and extended remote management from the smart meter. With AGI, real-time system information and control now integrates the functionality of advanced metering not only with demand response and outage management and restoration, but also with such distribution automation functionality as capacitor bank control, switch control, volt/VAR control, FDIR, fault detection and isolation and restoration, distribution management, and substation management.

Beyond integration of all these functions, AGI also includes integration at the DER level through inverter management, providing management of solar PV systems, EV charging systems management, and remote energy storage devices. To manage volt/VAR control of the new devices at the edge, the devices need the ability to make decisions on their own; intelligence at the edge is needed. The utility manages volt/VAR control at the substation level. The 6-kW EV power system will consume roughly the same amount of power in 8 hours of overnight charging as the house consumes in a 24-hour cycle. Such a dramatic increase in load will drive the need for edge power management. As described here, AGI functionality ensures that the utility will remain in control of all devices connected to the advanced smart grid, which has the added impact of addressing an emerging business risk for utilities when consumers gain increasing amounts of responsibility for edge device management.

Electricity and Smart Buildings and Appliances: Demand Response (DR)

More and more builders are adopting methodologies outlined by the U.S. Green Building Council (USGBC) and pursuing Leadership in Energy and Environmental Design (LEED) status. LEED is a classification scheme developed by the USGBC that provides guidelines and independent certification for builders who focus on unique design to promote not just efficient use of energy, but also more efficient use of water, lower CO_2 emissions, and improved internal comfort for the building occupants. Beyond LEED, electric utilities are promoting more efficient buildings through green building programs that encourage enhanced energy efficiency through new technology such as closed and open cell spray foam insulation and advanced window design, but also a plethora of less technological practices such as sealing around doors and windows, radiant barriers in attics, and blankets around water heaters, all of which serve to eliminate energy waste and reduce energy system capacity requirements, collectively postponing system demands to expand capacity by building new generation plants. Another aspect of energy efficiency concerns replacing appliances with newer, more efficient designs that consume less energy, most notably HVAC systems with higher SEER ratings, but also more efficient water heaters, refrigerators, CFL and LED lightbulbs, and so forth. As with smart buildings, smart appliances reduce energy consumption by substituting more energy efficient devices for less efficient ones. Finally, moving beyond the built infrastructure of smart buildings and smart appliances, *energy conservation* is rounded out when the focus shifts to a more active focus on human consumption behavior patterns throughout the day, leading to lower energy consumption throughout the day, but also to more fine-tuned conservation that can lower peak demand for optimal utility operations.

To draw the contrast between DR and energy efficiency further, changes in behavior designed to lower overall consumption fall under the *conservation* heading. However, changes focused on shifting consumption during specific times of day (peak periods when the costs to produce electricity are the highest) are referred to as *demand response,* when the demand side of the equation—energy consumers—respond to utility request to lower consumption temporarily to either reduce the production of high-cost energy (i.e., *economic* demand response) or relieve stress on an overburdened distribution grid (i.e., *reliability* demand response).

Looking 10 years into a mythical future, the scenario in the following use case imagines the impact rapidly changing technology will have on the electricity grid, driving the demand for an advanced smart grid and creating an alternative energy economy based on *negawatts*—avoided energy production.

Use Case of the Future: Demand Response (DR) in 2020

By 2020, most consumers were well on the path to shifting their electricity consumption away from peak periods, spurred by widely implemented DR programs coupled with new time-of-use (TOU) pricing incentives.

TOU rates required three key changes: (1) an advanced meter infrastructure system (AMI) with digital meters producing interval data, (2) a digital billing system that could produce bills that leveraged the interval data, and (3) an analytical study, followed by a TOU rate case.

Once in place, DR programs provided utilities three key benefits: (1) avoided capital expense from construction of new peaking power plants, (2) avoided operating expense from running aging power plants and from purchasing expensive power and transmission on the spot market during critical peak periods, and (3) enhanced profitability through selling relatively more power during lower-cost, off-peak periods.

By early 2015, utilities could begin to analyze the impact of the DR programs they had put in place. In effect, system operators had been seeking a new resource they could dispatch like traditional generation to keep the system in balance. Within 1 year of a DR program's implementation, megawatts of electricity demand would become available for load shedding, growing steadily as program acceptance grew. Increases in rates made negawatts from DR more attractive still, leading to hundreds of megawatts of DR capacity by 2020, effectively offsetting the construction of a midsized power plant for many utilities.

Another benefit of DR and TOU rates was that more and more demand shifted to match those times when renewable energy came on the system. In early DR programs, notification of curtailment need was manual, through phone calls or Web site postings. However, by the end of 2015, automation hardware and software could send digital signals to user home energy management systems (HEMSs), to user communication devices like smart phones and laptops, but also directly to appliances like smart HVAC units, water heaters, and refrigerators to prevent them from consuming power during peak periods.

By 2020, the DR system had proven itself sufficiently reliable and predictable to be built into integrated resource planning as just another resource to meet system demand. And together with efforts to improve energy efficiency, these DR programs enabled utilities to meet their portion of state mandates for peak load and carbon reductions at the lowest cost possible.

Power Engineering Concept Brief

As seen by the glimpse into the future offered in the use case above, DR holds great potential to become an integral part of the system of dispatchable resources under the control of the grid operator. Peak shaving, the earliest goal of DR programs, is a term used to define the reduction of the highest point along the load curve through targeted energy use curtailment. Peak shifting is a more complex form of DR that requires very close correlation with more sophisticated users, where the utility is able to move the peak and the load curve to more closely conform with its most efficient production curve. Carrying this concept still further, when the users are working in complete harmony with the energy producer, automated demand management controllers tailor the load to optimize the physical and economic delivery of energy to eliminate inefficiencies and maximize revenues. The ultimate for electric utility operations is a predictable, flat load curve, so that supply may be effortlessly aligned with demand.

To achieve this vision, HEMSs need to do more than just present detailed feedback on energy consumption to energy consumers. When equipped with automated algorithms, programmed to meet users' unique profiles of comfort, convenience, cost, and carbon (4C), the HEMS devices will be more effective at reducing loads in conformity with utility needs. When the HEMS devices are connected to smart meters through local wireless technologies like ZigBee, they will not only have access to interval data, but also provide an opportunity for utilities to engage directly with the HEMS devices when the customer allows it. Special pricing programs that can correlate with the 4C user profiles using the HEMS system will stimulate user behavior to more closely align with utility needs.

From a power engineering perspective, DR programs will first segment the market along a spectrum of the most willing, flexible energy users down to the least. These programs will present such groups with tailored tools, pricing, and programs that allow them to conveniently adjust their behaviors to meet utility needs. Beyond the DR programs, the tools will include HEMS (and for commercial customers, building energy management systems (BEMSs) that direct information to the device of the consumer's choosing. Those devices will need to correlate both with data received directly from inside the premises and with data from the smart meter and perhaps with data such as pricing information and special deals from the utility delivered through the smart meter (or over the Internet).

Interval data from smart meters, collected and organized with meter data management (MDM) tools, will be mined using data analytics programs to produce information that can feed such tailored DR programs. With the information and insights derived from the huge databases, patterns can be detected, causal relationships identified, and buyer groups formed to offer a new variety of energy services that appeal to consumers who grew up with one-size-fits-all commodity electricity.

Consumers and Prosumers: Distributed Generation (DG)

With DG, we see energy consumption converging with energy production. When energy consumers enable their premises to produce electricity using DG technology, be it with rooftop solar PV, microwind, combined heat and power (CHP), or other technologies, they create on-site power plants that come with an efficiency upgrade: they are not subject to line losses. Such a transition holds revolutionary potential, but is likely to evolve slowly for two principal reasons: first, the relatively high price of most types of DG when compared to grid-delivered power keeps DG solutions out of range for many consumers, and second, most energy consumers still lack the motivation or education needed to make them avid consumers of new DG solutions. But as technology progress makes DG both more productive and less expensive, as traditional electricity prices rise, and as DG becomes less exotic and more and more neighbors opt in, the appeal of going "off-grid" with a DG system will continue to grow.

A rooftop solar PV system is likely to be the predominate form of DG that a utility will encounter for the foreseeable future, given the flexibility and maturity of the platform relative to the other technologies. The principal components of a PV system include: (1) PV panels, (2) the DC/AC inverter, and (3) the net meter. For a PV system to become a DG node connected to the advanced smart grid, however, the inverter is the most likely component that will need to be made "smart" by adding a communication capability and localized intelligence. When that happens, the operator in the utility NOC will be able to "see" the asset and then send messages to control it. Such controls would provide direction that could use the energy produced by the smart inverter to achieve such system benefits as volt/VAR regulation and in peak times, they will provide access to renewable energy that could be marketed into the wholesale power market.

The challenges of connecting thousands, even tens of thousands of rooftop solar PV and other systems to the smart grid today remain manageable only as long as the amount of units per distribution feeder stays small, and as long as the energy each DG system produces doesn't flow unmanaged back onto the grid, which would put expensive distribution substation gear at risk—two significant constraints that could severely limit the uptake of DG as grid parity approaches. As with EVs in the next section, the challenge of integrating these DER systems is an issue that should be tackled sooner rather than later. Rather than pondering how many systems might be added to a utility service territory over the coming years before they must act, utility managers should be considering the threat that DG poses to equipment along a single, overburdened distribution feeder. Management at the feeder level will be critical to the integration and performance of DG in an advanced smart grid.

The term *high penetration PV (HPPV)* is used to describe efforts to load up a single distribution feeder with higher concentrations of PV facilities than current standards allow. A rule of thumb today is that a single distribution feeder can only handle about a 20% penetration of PV. Go north of 20% and the potential for intolerable risk and instability to grid operations sets a boundary—the intermittency of production and the potential for excess power to reverse the power flow on distribution feeder lines pose a threat to upstream equipment.

Another term, virtual power plant (VPP), describes a demand-side alternative to accommodate growth in peak demand to the traditional supply-side alternative of adding a natural gas power plant, commonly referred to as a *peaking unit* or a *peaker*. In its most expansive definition, a VPP combines an array of rooftop PV systems with localized energy storage and aggregated DR capacity (e.g., HEMS appliances equipped with direct load control—commonly, smart thermostats—or some combination). Such a system provides a utility the capacity needed to meet its peak needs without a power plant.

At the micro level, HPPV and VPP require technology to be refined at the scale of a single distribution feeder or neighborhood. Progress is afoot. Two innovative utilities, Sacramento Municipal Utility District (SMUD) and Hawaii Electric Company (HECO) are at the forefront of research on HPPV, conducting R&D projects in joint cooperation with the National Renewable Energy Labs (NREL), coordinating their efforts to provide valuable pioneer research [3] on the challenges and potential solutions regarding HPPV, using funds from the California Solar Institute and matching grants. Duke Energy is also testing VPP technology on a small scale at its McAlpine Creek Substation project [4], and the Center for the Commercialization of Electric Technologies (CCET) has a VPP project as a subset of its DOE ARRA Demonstration Grant project [5], where it will showcase its Smart Grid Residential Community of the Future in a suburb 25 miles north of Houston.

Use Case of the Future: Distributed Generation (DG) in 2020

After the conclusion of national elections in late 2016, when Congress passed a landmark climate bill that mandated 85% carbon reductions by 2050, electricity leaders nationwide knew they would need every clean energy resource they could find, which spurred a boom in renewable energy. Small-scale distributed generation (DG), in many ways, was well suited to the challenge of climate control.

DG put residential and commercial energy consumers in charge of their own destiny, with greater self-sufficiency and security, insurance against future energy price shocks and a chance to take advantage of major new economic opportunities, including thousands of local jobs. The transition to DG started simple enough, but that is not to say that there weren't bumps in the road. Most of the solar PV capacity would end up coming from crystalline silicon and thin film solar photovoltaic modules, the leading technology options at the time. Installations were mostly on residential, commercial, industrial, and civic rooftops, but some small, local solar farms were also built, many of them pulling double duty as parking lot shades and charging stations for electric vehicles.

The benefits of DG were broadly shared: besides all the homeowners who installed rooftop systems, residents without rooftops or backyards were able to buy shares in midsized solar cooperative farms (500 kW–2 MW). Utilities also rented large commercial building rooftops to deploy utility-owned solar, leased solar modules (and near the end of the decade, energy storage systems), and bought local clean energy from private developers, packaging DG with low-carbon centralized generation to expand affordable clean energy.

New energy storage technologies, as they came on line in the second half of the decade, did more than anything else to spur greater interest in DG, transforming intermittent renewable generation into firm, dispatchable power. That allowed a utility to buy and store low cost energy to use at peak, high-cost times. The list of utility benefits of stored DG included the ability to: (1) replace spinning reserve, purchased power or new peaking plants, (2) defer investment in T&D upgrades, and (3) improve power quality, reliability, and outage management programs.

While most discussion about DG concerned solar PV, it should be noted that solar PV did not tell the whole DG story. An economically compelling slice of DG also came from such technologies as combined heat and power (CHP) and the conversion of landfill gas (LFG), waste heat, and waste biomass into electricity. These relatively low cost opportunities were already at or below parity with fossil fuel generation at the end of the first decade of the new century. While the potential of waste heat and biomass was limited to how much waste and biomass was available, such baseload technologies proved uniquely valuable for reducing greenhouse gases. First, every lump of coal avoided because of waste heating and energy production also meant avoided methane emissions, and since methane has more than 20 times the global warming impact of carbon dioxide, waste energy became more and more popular. By 2020, the conversion of waste to energy has become widely valued as an important component for a utility to achieve its environmental goals.

Power Engineering Concept Brief

In power engineering terms, what is a smart inverter and what is the required functionality to make a solar PV system a dispatchable node on an advanced smart grid? First, let's discuss the current design and functionality of an inverter connected to a standalone solar PV system. The fixed output inverter is one of the simplest designs possible for an inverter, and the inverter most often chosen for a solar PV system is known as a true sine wave (TSW) inverter, which produces a true sine wave as the name suggests, providing high-quality power that does not produce adverse effects. These inverters have a principal task: convert the DC power output of the solar PV system into a steady stream of AC power that can run through a net meter to be measured, then on through the circuit box to be consumed within the premises. If the load inside the building is less than the output of the system, the system sheds the excess power to ground. In effect, the tasks demanded of this inverter are relatively straightforward compared to what we will ask of the smart inverter–equipped solar PV system. For the purposes of this analysis of the smart inverter, let us also assume that the local distributed generation system includes a local energy storage unit to provide more flexible use of the power produced by the smart solar PV system.

The smart inverter is equipped with two key capacities that separate it from its simpler cousin. First, the smart inverter possesses localized intelligence—business rules embedded on a chip—that lets it make decisions that suit both the needs of the system owner and the needs of the utility. Second, the smart inverter is also equipped with communication capability, either to communicate locally with the nearby smart meter, or to communicate over longer distances within a local or regional network. Ideally, a third key capacity will be added to any smart inverter in the future—power electronics that enable the output of the smart inverter to be tailored to the needs of the system, varying voltage to VAR output as needed.

In one way, the smart inverter acts as a dispatching agent, deciding on a moment by moment basis whether the power should be stored locally, fed to the premises to offset grid power consumption, or fed back into the grid. To make such decisions, the smart inverter requires information on the current and historic power consumption at the premises, the charge state of the connected storage device, and the market price and grid's ability at any time to accept voltage or VARs onto the grid.

The advanced smart grid is capable of managing thousands of these new smart inverters, automatically dispatching their power when needed to optimize the grid and to take advantage of market opportunities where renewable energy is priced at a premium. Having power input both from centralized generation at one end of the network and thousands of smaller distributed generation units at the other end provides a radical new capability to grid operators to achieve grid optimization.

Electricity and Transportation: Electric Vehicles (EV)

Transportation and electricity infrastructure have much in common, but historically the transportation industry has been dominated by a dependence on petroleum products, which has resulted in electric vehicles (EVs) being relegated to tight niche purposes, such as mass transit vehicles, airport vehicles, and forklifts in industrial facilities. There has been little chance of the two industries converging when it comes to personal transportation. But considerable progress over the past decade in the technology, design, and development of a commercially viable EV has changed that, leading to an explosion of interest and the need for all electric utilities to reconsider the potential impact of a massive EV adoption, especially given the impact on specific distribution feeders in potential high adoption areas of a distribution grid. Convergence of these two foundational industries is now not just likely, but a foregone conclusion. Before moving on, a word or two is in order on EV and charging station technology.

While battery alternatives for EVs range from lead acid to nickel metal hydride, the technology with buzz appears to be lithium ion, more specifically, lithium ion phosphate, which seems to offer the most appealing combination of low weight, high specific energy, and energy stability to make it the most appropriate technology selection for transportation applications. Regardless of the battery technology, however, electricity capacity in a battery is measured in amp hours, with total energy capacity denoted in watt hours. Charging an EV depends on the voltage at the socket and in the charger capacity, both external to the vehicle and in the on-board charger itself. As stated previously, a good rule of thumb is that a typical EV is likely to double the electricity consumption of a typical home. Clearly, the impact of this huge jump in electricity consumption on a localized basis in a distribution grid must be considered by electric utilities well before EVs become widely adopted.

EV charging station vendor Couloumb is leading a program called ChargePoint America [6], which provides a lesson in the enormity of the task in establishing a charging station infrastructure. Announced in June 2010, the program has plans to deploy 4,600 public charging stations in nine regions around the country by September 2011 (to put things in perspective, about 115,000 gas stations comprise our current gasoline-powered vehicle infrastructure). The program is supported by the DOE's Transportation Electrification Initiative [7] and a $37 million budget, which includes a $15 million ARRA grant. While there may ultimately be commercial charge stations that resemble today's gas stations (see Better Place [8]), EV charging is different in that it can occur throughout the existing electricity infrastructure, in homes, at businesses, in parking garages, on parking lots and at curbside, wherever a plug is found. It is not hard to imagine charging stations being used to attract customers to businesses—the Whole Foods grocery store in Austin, for instance, will likely be one of the first to deploy charging stations in prime parking spots.

The rate of EV charging is measured by the voltage used over time by an EV charging station. *Level I* charging occurs at the standard voltage of a typical electrical outlet in the United States: 110 to 120 volts, which can result in a charge period of between 8 and 16 hours. *Level II* charging is more suited to overnight charging, taking from 4 to 6 hours at 220 to 240 volts (the voltage typically found in the outlet used for a clothes dryer, which is the rate offered by ChargePoint America). When away from home base, EVs will seek a more rapid charge, so *Level III* charging uses 440 volts, providing an 80% charge in as little as 30 minutes. But Level III chargers are more expensive and given the high voltage level, require substantial training and facilities.

In contrast to the charging model, Better Place [8] has adopted a charging infrastructure approach similar to the gas station (more accurately, a Better Place charging station would resemble a gas station combined with a car wash). The Better Place business model actually swaps the entire depleted battery pack for a fully charged one, using an automated line with a robotic arm that detaches and reattaches the entire battery assembly underneath the car body in a matter of minutes, while the driver waits. The customer enjoys a much more rapid charge experience compared to the Coulomb approach above.

Rather than paying for the battery pack as part of the vehicle, the customer pays for the energy consumed, but does not own (or take the risk of owning) the battery pack. This model has so far gained traction only in ecosystems of limited size: the small countries Israel and Denmark came first, followed by the island state of Hawaii. Given the expense of building these charging stations, this model may or may not be limited to such small ecosystems, only time will tell. But Better Place appears to be a viable alternative to the charging infrastructure and business model outlined by Coulomb's ChargePoint America.

Three key elements provide insight into the convergence of transportation and electric grids: (1) EVs as new electricity demand, (2) EV charging station infrastructure, and (3) EVs as a utility energy storage resource.

First, EVs represent *new electricity demand*. Because the fuel source of an EV is electricity, these vehicles represent a significant new load to be added to the electricity grid. The impact of an EV on grid operations depends on the rate of EV charge, the frequency of charging, the time of the charge (peak, off peak, and so forth), the location of the charging, and the level of charging coordination (planned, unplanned, and so forth). The key question to answer concerns how to manage this new type of load as a potential burden or threat to utility operations. If electric utilities can shift this new load to times of day when the utility's generation resources operate with considerable slack, then they can achieve greater capacity factors and efficiencies and improve profitability. But if EV owners plug in when they arrive home, the collective new EV load during the peak periods in the evenings and during the summer will stress the grid still further and require significant capital investment to make the grid more robust.

Second, as discussed above, EVs require a *charging infrastructure*, likely comprised of three main charging alternatives: (1) *on-premise* charging stations for the "home" location, whether a residence or a business (Level I or II), (2) *public* charging stations, open to transient EVs on a scheduled or ad hoc basis (Level II or III), and (3) *private* charging stations, closed to public use but open to fleet EVs (Level II or III). The design of the infrastructure will be a critical issue for utility operations, as it will go a long way to determine how and when EVs interface with the grid. The key question to answer concerns the optimal infrastructure design for a utility.

Integrating the emerging EV infrastructure into the utility grid, into local communities and into individual households will require tremendous cooperation and planning. Utilities will need to roll up their sleeves on this one—they will need numerous trials to refine the details and determine the appropriate *vehicle-to-grid* (V2G) and *vehicle-to-home* (V2H) processes and policies.

Finally, EVs represent a potential new *storage resource*, if one that remains a long way off. The challenge in this regard will be to determine the best way to take advantage of a fleet of distributed storage elements with a significant amount of collective storage capacity. As consumer adoption of EVs progresses, the collective storage capacity of fleet vehicles will grow as well, giving promise for a new clean-tech resource for utility-wide load shifting. Perhaps the most intriguing proposition regarding EVs is that they could be used to store renewable energy production, particularly power from wind farms, which produce most prolifically during off-peak hours (i.e., during the night), when energy demand and energy prices are lowest. Using EVs to store wind power during off-peak periods could provide significant arbitrage value if that stored energy could be discharged to the grid during peak periods when both demand and prices were far greater. The key question to answer concerns how to leverage this new type of resource as a potential opportunity for utility operations. A burgeoning V2G storage capacity will enable smoother utility distribution system operations, and a growing V2H capacity can be expected to provide a significant boost to the potential of DR programs discussed above as well.

Use Case of the Future: Electric Vehicles in 2020

By 2020, the signs of the electrification of transportation systems were increasingly apparent. Roughly 35% of new car purchases were electric vehicles (EVs), a category that included not just plug-in hybrids, but all electric vehicles. Incentive programs engaged utility customers and EV dealerships to install the charging station infrastructure to help support the surge in EV ownership. New EV owners, both individuals and fleets, were drawn by the combination of high fuel costs and popular incentives. For a city of 1 million, these penetration numbers meant that about 200,000 EVs circulated daily throughout the city in 2020.

Electrification wasn't just about EVs, though. Mass transit became more electrified—legacy rail lines used electricity as a growing substitute for diesel-electric rolling stock. Preliminary deployments of EV bus fleets were underway, and in airports across the country, full electrification of ground support equipment had been completed for years.

EV program planners discovered more usage patterns than the nightly charging scenario that took advantage of low rates and coincidence with wind energy production. Some users had to recharge during the day based on their personal schedules; others participated in EV car-share programs, which required frequent recharging between short drives. School buses faced heavy use for nine months, but little to none during the summer, freeing their storage capacity for other purposes. Some workers had chargers at their workplace, and some multifamily housing developments and retail establishments added charge stations to make their locations more competitive. The EV charging market developed steadily as third-party companies sited recharging stations throughout the city at strategic locations, most especially high-traffic retail sites, mass transit parking lots, and on the rooftops of parking structures. Finally, city building codes needed to be adjusted to accommodate EV charging, reflecting the more integrated approach adopted by utilities over the course of the decade.

As utilities integrated EV charging capabilities with their smart grids, they gained a new automated energy storage resource. Charging and discharging were programmed over the smart grid to achieve an optimal balance between grid needs and the needs of the individual EV owner. Dubbed electric vehicle support equipment (EVSE), these smart chargers accommodated the new time-of-use (TOU) rates and fed relevant information to digital billing systems when and as needed, including to support TOU rates, real-time pricing (RTP) options, generation fuel-mix forecasts, wind generation signals, and other customer-specific information, providing a range of charging options (e.g., charge at lowest cost, charge for lowest carbon footprint, charge immediately). The EVSE was designed on industry standards in cooperation with other utilities and the EV industry to ensure interoperability, flexibility regarding communication types and low cost production. In 2020, the relationship between the consumer and the utility had evolved, but the complexity was managed by ever more capable technology and automation.

Power Engineering Concept Brief

EV as Stationary Load

In the utility world, load or system demand needs to be managed and planned, and that makes the unpredictable nature of EV adoption rates a huge challenge.

For the sake of clarity, let us avoid the issue of mobile power consuming devices and focus here on the management of stationary charging stations and the load they will place on the utility, whether they are located in a residential customer's garage, on the curb, or in the parking lot of a small business. If for no other reason than to ensure that the utility will not take an unplanned risk with regard to mobile EVs, it is essential that system planning to manage charging station loads be coordinated with these different customer classes. Utilities will need to develop a strategy that targets specific distribution feeders for upgrades, to harden them against the risks of overloading from too many EVs charging at once.

EVs as Roaming Load

Mobile load represents a unique management challenge to a utility, never before encountered given that providing power to meet demand has heretofore been relatively predictable with regard to any grouping of grid termination points. Thus, utilities may also consider the relative costs and benefits of two strategies seemingly at two ends of a spectrum of options: either corralling and clustering EVs when they are roaming by concentrating charging stations, which would be a strategy of providing charging stations to garages and parking lots, or distributing and dispersing charging stations throughout the territory to defray the impact they may have on the system at any one time. Of course, planners may choose to do one strategy before the other or to do both simultaneously. The idea is to design a charging system infrastructure that drives EV charging activity to an outcome that best conforms to the needs of the utility.

Examining how utilities have managed the deployment of rooftop solar PV systems is instructive to discern how they might approach EVs and related charging infrastructure. Some utilities have let the market control where solar PV systems are deployed, perhaps offering rebates to defray costs, perhaps under rules and guidelines for system connectivity (i.e., Net Meter), but in essence taking a hands-off approach. Some utilities may choose that path for EVs, perhaps predicting that the pace of change will be slow enough to manage the situation and learn from the outcomes. The benefits of this approach are to allow the market to grow at its natural pace, in the vicinities that it may, and to help define the future infrastructure investments that match the market requirements. The costs are lagging the adoption of the EVs, so that it becomes a deferred investment strategy. If the utility is not proactive, however, the utility is by default forced into a reactive stance, unable to manage the EV-related peak load, bringing on instability.

In other cases, the utilities will take a more hands-on approach by proactively locating community solar PV systems, investing in rooftop leasing programs, educating the public on the dos and don'ts of solar PV, and enabling new market opportunities to emerge, in essence partnering with the market to help control outcomes. Similarly, some utilities may do the same thing with EVs.

The benefits of this approach include proactive planning for the location and adoption rate of EVs. The costs will include new EV charging stations across the planned locations, new power electronics to accommodate new load concentration, and new customer support programs.

The disconnect today is that current EVs adhere to certain standards in which the utilities also participate, but the first generation EVs are not yet integrated as dispatchable assets to come under the control of the utility, so that the utility is not able to control with any specificity when and where the EVs charge. Utility investment in charging infrastructure then becomes a strategy to take control and manage outcomes. In this approach, the utility subsidizes the charging equipment, not unlike the cable companies subsidize the DVR, both to generate ancillary revenue and for operational control. Issues that the utility needs to address in managing a charging infrastructure may include location, design, and functionality of the charging stations, customer identification, metering and billing, and the new power electronics needed to enable such functionality and control.

Imagine a group of friends driving to Austin, Texas, to attend the 2015 South by Southwest [9] (SXSW) Interactive, Film, and Music Festivals during the spring season. They drive down from Seattle, Washington (in their new "extended range" EVs). As the driver sits down, she types into the navigational system "SXSW in Austin." The EV maps the best possible route based on the driver's choices (e.g., cheapest charging, cheapest hotels, best restaurants, and best scenery). As the friends start their journey, the EV has already contacted Austin Energy (utility in Austin, Texas) to open an account (or reactivate an old one), providing the driver's information (name, address, telephone, credit information, and so forth), and prenegotiating the best rates and locations to charge the EV while in town, based on the driver's preset condition parameters.

While the driver and her friends are parked at SXSW enjoying the festivities, the car becomes a provider of energy and energy services to the grid, generating revenue for the driver, (hopefully sufficient to offset parking charges!) The car serves as an on-demand capacitor to the local grid via the "SXSW-EV Program," which pays a premium above regular rates during shoulder peak and full peak hours to ensure that the utility can manage the situation and to gain benefit from the influx of EVs, which are in effect mobile storage units—DER devices.

Electricity and Warehousing: Energy Storage

At the risk of stretching this convergence meme as far as it will go, we come to the end of this chapter with a discussion on the introduction of energy storage (ES) into the electricity supply chain. Integrating energy storage into the electricity distribution supply chain must be viewed as the "mother of all game changers." Consider that throughout the supply chain, the rules of the game have forever been written around the fact that the system must operate in *real time* because of the *lack of economic energy storage alternatives*. There has never been much of a warehouse alternative in this particular supply chain. Having only hydropower as a feasible, widely adopted "warehouse" alternative, the electricity supply chain developed as if energy storage would always be: (1) very expensive, (2) difficult to site, ruling it out in most instances, and (3) rather clumsy in its application, not providing the fine tuning one would hope for in a storage asset. Therefore, it should come as no surprise that the electric system developed as it did over the next 100 years, designed and operated according to such assumptions as: (1) the system must be kept in balance, (2) most load is unchangeable, so we must get good at following load with generation to keep the system in balance, (3) the primary challenge is to manage the system capacity to ensure availability during peak consumption periods, and (4) supply-side resources are the dominant solution to add significant capacity to the system.

Where grid managers and designers could employ hydropower as an energy resource, they did. The first significant power plant was a hydroelectric dam between Niagara Falls and Buffalo, New York. Indeed the early history of electricity is very much concerned with acquiring rights to rivers and building hydroelectric dams. Hydropower, in fact, has subsequently been adapted into a particular energy storage technology referred to as "pumped hydro," where smaller paired reservoirs are constructed for the specific purpose of energy storage—using electricity to pump water up when electricity is cheap, then letting gravity drop the water through the system to generate electricity when market prices are elevated.

Early adoption of energy storage is seen in deployments that facilitate the addition of variable renewable energy resources onto the relatively small transmission grids of the Hawaiian Islands. We also see energy storage proposed for the intersection of the three major grids in the United States at the new Tres Amigas project in New Mexico [10]. Energy storage is also being deployed to provide ancillary services on the NY and PJM grids and in a research project in South Texas.

Use Case of the Future: Energy Storage (ES) in 2020

For those utilities that had added utility-scale storage to their distribution grids, the utility operational model that had worked for over 100 years had been turned upside down. By 2020, several storage technologies had reached well into commercialization stage and price points were coming down, though prices still remained too high for many utilities. Even in 2020, many utilities remained paralyzed by the diversity of the storage technology options available and the rapid changes.

Those utilities that did invest in energy storage had more options, because now energy could be economically stored and used when it had greater value and more utility. Finally, like nearly every other industry, the electric industry had a warehouse capability. How did they use it? Clearly, utilities were still in the experimental phase. The 2009 ARRA legislation, more commonly referred to as the Stimulus Bill, had primed the pump back in 2009, when billions flowed through the DOE into utility programs, especially the storage demonstration projects under DOE FOA 36 [11].

Leading companies emerged for the different types of energy storage, and clear leaders for each type of application emerged as well. Managers at utilities had their preferences. Some technologies were simply better for some applications. Utilities had moved slowly into energy storage, but once a utility found one or more applications and technologies that were market ready and fit their needs, they made rapid progress. In the early days, utilities used pilots to investigate the possibilities, from combining smaller energy storage systems—"community energy storage"—with solar PV panels in a neighborhood to placing larger, utility-scale energy storage facilities on industrial sites for load shifting to avoid peak consumption, to using energy storage to relieve congestion at strategic points in the grid, to collocating energy storage with renewable energy farms to provide a buffer against disruptive intermittent power production.

Power Engineering Concept Brief

In this section, we are more concerned with energy storage as it converges with the role of warehousing in a supply chain rather than with individual energy storage technologies. Our focus is thus on the supply chain impacts of energy storage and on the smart inverter, the point of connection between the storage technology and the advanced smart grid. In considering energy storage from the power engineering perspective, in the near term the price of energy storage will keep storage devices as a precious commodity, so the choice on where to deploy such an asset will likely be driven by where it is most likely to provide the most value, whether it is to accomplish a business goal or to provide information and insight in a pilot or research project.

Energy storage has great potential as an element to transform the design of the advanced smart grid. Energy storage devices will need to be integrated into the advanced smart grid with the following issues in mind. At first, integration will be accomplished in phases, with energy storage systems added incrementally while energy storage technologies become more and more economically feasible with dropping prices. Thus locations on the grid will be targeted based on some combination of economic and system engineering benefits. Second, energy storage will be considered as a critical element in disaster recovery, which indicates collocation with shelters and critical facilities. Finally, energy storage will accelerate the processes and changes described in this chapter—when energy storage is added, the different components, applications, and design elements of the advanced smart grid described in this chapter become more efficient and versatile, from DG to DR, from EV charging stations to DA. Energy storage, the most versatile technology when it comes to grid operations and enhancements, reshapes the potential of the grid.

Imagine for a moment the load management strategies that grid operators could achieve if they were to deploy energy storage devices across their service territory to accelerate outage restoration times, minimize load congestion zones, optimize large disaster restoration zones (e.g., schools and churches), improve small disaster recovery zones, and improve all around volt/VAR control, all while improving quality performance indices such as SAIDI, SAIFI, and CAIDI.

In particular, let us look at one kind of energy storage. Thermal energy storage has hidden potential, as described herein. Refrigerators and freezers, ubiquitous in households and many businesses, have a primary role to store cold air to keep food fresh. Looked at a different way though, these devices are also microwarehouses of thermal energy, which when aggregated, may serve as a resource on the advanced smart grid. During peak periods, such connected devices may be signaled to switch to a conservation mode that will postpone their regular chilling cycle, so that they become a distributed resource not currently in play on the grid. These distributed thermal energy storage devices change the way we look at storage and appliances. Integrating such assets need not be about any loss of comfort or convenience either, rather tapping such a resource merely involves a minor sharing of a thermal storage resource in a collective strategy to incorporate a new resource that did not exist before. And the advanced smart grid will make this possible by providing a network, a network management system and the smart devices—the smart inverters—as the missing elements to access distributed elements and realize their hidden value.

The key to realizing these and other scenarios lies in the smart inverter, which when connected to the distributed energy storage device transforms it into a smart grid element. The smart inverter, like the smart meter and the smart router, has local intelligence thanks to its processor, and communications capabilities (in silicon chips that provide for Ethernet/LAN/WAN connectivity).

Conclusion

In this chapter, we highlighted two megatrends that are transforming infrastructures as diverse as the electric grid and the state highway system: digitization and networking. In fact, all infrastructures have the opportunity today to add networked digital sensor devices to gather information on infrastructure status and operations, whether the commodity they move is electrons or vehicles. In short, all infrastructures benefit not only from access to such revolutionary digital technologies, with new devices emerging every day, but also from advances in network technologies that enable that information to flow back to management consoles, databases and servers, where the data is acted upon with new data analytic software to provide insights never before available. These trends make formerly diverse infrastructures more alike and lead them to work more closely together, even interoperating in some cases, as when water systems automatically curtail their usage during peak electricity periods to save the energy that would be used to pump and treat water during those critical times. We've labeled these activities using the term *smart convergence*, where the infrastructure managers learn from each other and where possible, leverage each other's infrastructures to achieve still greater operational efficiencies.

Infrastructures that benefit from smart convergence share the following core functions: (1) *distributed*, which have elements they draw upon throughout the infrastructure, (2) *interactive*, where the elements of the infrastructure interoperate and influence each other, (3) *self-healing*, where the elements work together in such a way as to promote improved performance, and, finally, (4) *ubiquitous*, in which their converged qualities are found in every device.

Smart convergence, as it is recognized and employed by infrastructure operators, will have significant consequences for all aspects of our modern economy, which rises and falls based on the success and health of its multiple infrastructures. Smart convergence will lead to dramatic cost reductions when infrastructures share common elements, but also to dramatic increases in effectiveness when a combination of infrastructures leverage efficiencies or when borrowing best practices, improving operations and changing potential by recognizing and incorporating new assets heretofore unavailable.

In Chapter 4, we will trace the origins of the advanced smart grid concept in an extended case study of Austin Energy, starting in 2003 with reforms taken to make the IT back office more efficient and cut costs, leading through a variety of projects to address application silos and integrate new applications and smart devices, and finally resulting in the emergence of a pioneer utility-wide smart grid in 2009. The 7-year process proved both valuable and instructive, achieving its stated goals, but also revealing lessons learned and insights on smart grid through both its successes and failures.

Endnotes

[1]   http://www.utc.org/.

[2]   http://www.thocp.net/hardware/atm.htm.

[3]   http://www.calsolarresearch.ca.gov/Funded-Projects/solicitation1-smud.html.

[4]   http://www.duke-energy.com/news/releases/2009061602.asp.

[5]   http://www.electrictechnologycenter.com/doe.html.

[6]   http://www.chargepointamerica.com/.

[7]   http://apps1.eere.energy.gov/news/daily.cfm/hp_news_id=159.

[8]   http://www.betterplace.com/.

[9]   http://sxsw.com/.

[10] http://www.tresamigasllc.com/.

[11] http://www.energy.gov/recovery/documents/xDE-FOA-00000.36.pdf.

4

Smart Grid 1.0 Emerges

Chapter 3 explored the different infrastructures that we depend upon for our modern economies and lifestyles and showed how they are converging on themselves. The electricity grid in particular is incorporating telecom practices and habits, and IT is becoming an ever more vital aspect of electricity grid operations. In Chapter 4 we use a case study to showcase and evaluate such concepts in a single utility, examining in detail an actual Smart Grid 1.0 implementation. The case study approach reveals any number of lessons learned, but also details the emergence of the advanced smart grid vision, showing how it developed through trial and error in a real-world living laboratory environment.

Introduction

This chapter describes the genesis and implementation of a smart grid at Austin Energy, the ninth-largest city-owned electric utility. Starting in 2003 when the term "smart grid" had barely been circulated among utility cognoscenti, and had certainly not gained the widespread acceptance it enjoys today, we defined what is today called a smart grid through the series of incremental steps we took at Austin Energy (AE) described in this chapter. The smart grid as it came to be defined at AE was more expansive than other definitions at the time (EPRI's Intelligrid, IBM's Intelligent Utility Network, Meta's Geodesic Energy Network, and so forth), which were centered on the utility electric infrastructure alone. Starting in 2004, when Austin Energy's CIO Andres Carvallo first started talking publicly about a smart grid, he combined the electricity infrastructure owned by the utility on one side of the meter, with infrastructure beyond the meter owned by customers, from private and publicly owned buildings to individually owned electric vehicles, to distributed energy resources and smart appliances.

This chapter describes the vital first step of technology infrastructure rationalization, critical to the future success of any smart grid project. It explores the creation and adoption of a technology governance framework and a smart grid architecture, as well as the expansion of networking assets (fiber to every substation, wireless AMI system deployment territory-wide, wireless demand response system extended to smart thermostats, wireless distribution automation system reaching a limited number of sensors for grid optimization), and the addition of a variety of applications through over 150 separate systems integration projects to create the first smart grid. Finally, the chapter describes how all this work led to the insights for a new approach to building a smart grid, which we have labeled an advanced smart grid as described in the first three chapters.

As will be demonstrated by the case study below, there were two key lessons learned that deserve highlighting. The first one concerns the importance of preparing the IT department as a firm foundation for the smart grid to follow. Without a rationalized technology infrastructure, the complexity of a smart grid project guarantees substantially higher risks and costs, even so high as to put the project itself at risk. The second one concerns the revelation that beginning a smart grid project by acquiring network capabilities, instead of by incremental addition of applications, provides immeasurable benefits, as described in the first three chapters of this book.

The authors of this book, Andres Carvallo and John Cooper, collaborated twice in projects at Austin Energy as it was building the nation's first smart grid deployed over a full utility environment. Andres Carvallo served as CIO of Austin Energy from 2003 to 2010 and was the visionary, principal executive champion, and chief architect of Austin Energy's Smart Grid. Andres Carvallo hired John Cooper to work as a utility applications and IP network communications consultant for Austin Energy in 2004, and then later in 2009 as a project manager of the smart grid team within the Pecan Street Project, described in full detail in Chapter 5.

Case Study: Austin Energy, Pioneer First Generation Smart Grid

This chapter tells the story of how the very first comprehensive, utility-wide first generation smart grid came to be built in Austin, Texas, at Austin Energy, the city-owned electric utility that serves over 1.2 million residential end users and 43,000 industrial and commercial businesses, distributing electricity to over 410,000 meters. The lessons learned in the smart grid journey described in this case study are fundamental to understanding the concepts in this book, which derive not only from the successes and lessons learned in Austin, but also from our work in Texas and in the United States from 2003 to 2010.

To enhance clarity and provide greater insight, we present the remainder of this chapter in a case study format, as told from the perspective of the CIO during that time, Andres Carvallo.

Saying Yes to Opportunity

Juan Garza, the general manager at Austin Energy (AE), sought my help in January 2003, to leverage technology to transform the city-owned utility into a new kind of organization capable of greater flexibility, so it could adapt to the changes that were sure to come. A future journey of personal and organizational transformation began with a phone call requesting an interview (Juan had heard I was recently back on the job market). Juan told me about his ongoing search for a leader to manage IT inside the utility, describing a difficult situation: AE was not efficient in the use of high-tech resources, Juan told me, as he detailed the lack of tools, reports, and visibility that frustrated executive decision-making and management of the enterprise. Juan and his executive team needed better processes and systems and better integration between IT and other parts of the utility.

That morning, Juan described to me a utility that is not unlike many in the developed world today. The information technology and telecommunications (ITT) department was responsive to the other departments at the utility, but its best efforts were too often stymied by events and constraints seemingly beyond their control. The budget was forever inadequate—demand exceeded supply. More than 90% of any annual IT budget was used to maintain current systems (also known as "keep the lights on"), leaving little to no money to finance the strategic projects that promised to lift the department out of its daily frustrations and deliver the capabilities sought by the other business units that were clients of the ITT department. Juan described an ITT organization that was not working well, providing little information for executives to manage the enterprise.

Before going further, Juan asked me to meet with some of his staff to get some perspective. I met with several of his direct reports, including AE's CFO Elaine Hart and Roger Duncan, who oversaw all government relations, renewable energy, and energy efficiency programs at the time. I asked them both a series of questions about their use of technology, reports, and systems, how they ran their businesses, and so on. These one-on-one sessions helped me get a better idea of what was happening and I returned to share my assessment with Juan. Citing an array of problems, I confirmed the challenges Juan had described to me when we met—the aging infrastructure, inadequate systems, lack of project management, lack of enterprise architecture, lack of even a single version of the truth, and no key performance indicators (KPIs) used for benchmarking performance.

Juan asked me for a price tag and a time frame estimate to fix the enterprise, and I gave him a ball park estimate of spending an additional $50 million, over and above existing technology operating and capital expense budgets, to save $100 million. The central concept we discussed was to invest more in technology to achieve savings in capital and operations expenses that would provide sustainable efficiencies across the enterprise. The annual IT budget increased significantly during my tenure, but the corporate numbers tell a more meaningful story. When I joined AE in 2003, we had about 1,500 employees, 320,000 meters under management, annual revenues of about $750 million, and no online services of any kind. By the time of my departure in early 2010, staffing had grown to about 1,700 employees—about 12% increase from 2003—but customers and annual revenues had grown comparatively more, to 410,000 meters and about $1.2 billion in annual revenue—about 30%, and smart grid plus full online services had been built.

This then is the story of the smart grid in a nutshell: how we used technology to improve services and enable lean growth. This case study captures the success of a pioneer smart grid journey I was fortunate to have led.

A Fresh Start

In 2003, Juan asked, "Well, do you want the job?" "What is it we're talking about?" I countered. "You come run technology, and we'll invest to free up OpEx and CapEx. We're moving towards these new concepts—distributed generation, energy efficiency, even electric vehicles, and we need to get our house in order."

I accepted the offer, and on February 18, 2003, I walked into my new office for the first time, when I was introduced to a strategic team of executives and senior managers. From the outset, we worked together as a team to redefine the values of the company, including a new mission statement: "To deliver, clean, reliable, affordable energy and excellent customer service." I was fortunate, as I said, to come in at a time where there was significant transformation already underway and technology was accepted as a means to an end, helping to integrate new concepts, systems, and processes.

In their utility vision, Juan and Roger saw a bright line between an old way of doing business and a new way. A couple of years into my new job, Juan's vision was finally realized with the completion of a reorganization around two divisions, with Bob Kahn as Deputy GM of Administrative Services and Roger Duncan as Deputy GM of Distributed Energy Services. A few years later, Bob left to become the CEO of ERCOT, the independent system operator for Texas, and Mike McCluskey stepped up to become the deputy GM of Centralized

Energy Services and chief operating officer. In 2008, Roger would succeed Juan as GM, when Juan left to lead our neighboring utility, Pedernales Electric Co-operative. During the transition, most of the resources remained on the central-ized side of the house, but the distributed side grew rapidly, playing catch-up as it integrated new systems and technologies.

As for the ITT department, we had the task of working with both halves of the utility to ensure reliability and continuity of service for current business operations and seamless integration of new functions to the degree possible. In important ways, the ITT department helped to lead the utility in its transfor-mation while I was there, but in other ways, as I said earlier, the transformation to a "utility of the future" vision was already well underway and our department had to keep up with my innovative colleagues to enable such landmark nation-leading programs they were rolling out, such as GreenChoice, the green energy power purchase program, the Green Building program, and the PowerSaver Free Thermostat program.

Initial Assessment and Issue Identification

I set to work in the first 30 days with an initial assessment. I met with my staff and began interviews to understand technology service delivery and challenges. I also met with a variety of line managers on the operations side—ITT's inter-nal customers—the better to understand their roles and how they used tech-nology and IT services. Those interviews began with owners of customer care, marketing, and finance, and soon thereafter, with owners of operations technol-ogy (OT) including electric service delivery, power generation, and wholesale trading.

At this point, it's worth a pause to compare OT and IT, because this relationship is critical to the success of any smart grid transformation. We're all very familiar with IT, but the term OT has less circulation. In the utility, OT describes the power engineering and operational technology groups that manage the generation, transmission, and distribution of power. OT generally is used to refer to wholesale trading operations, as well as power plant engi-neers, the T&D department, line crews, and metering teams. In a smart grid transformation, it is critical that the IT staff engage seamlessly with the OT staff to upgrade and transform the energy platform to include new information and communication technologies and process innovation. In fact, I would now venture to suggest that OT and IT would be better located together under one executive to truly achieve success.

After those initial meetings, I decided a deeper assessment was in order, so I picked two business analysts in my team to help document the assessment. We set up about 500 one-on-one interviews (out of close to 1,500 total employees) and interviewed each one using a common questionnaire, asking questions such as "What do you like/dislike about technology? What works/doesn't work well? What systems do you use? How well do they work? Is there a replacement strategy for those systems? How do you get support? What happens when things break? Are the systems internal or customer-facing?"

The results of that questionnaire provided data that let us produce a systems inventory of all technologies currently in use, from the perspective of the users. As we analyzed the interview responses, it soon became clear what was working and what wasn't. As Juan had suspected, our analysis identified a number of fundamental problems.

Issues identified in the initial assessment fell in three major categories. First, the requirements on the ITT department were out of balance with the available resources, principally due to the large number of legacy programs and a lack of coordination and synergy. Second, the necessary IT tools and processes to run a first-class organization were lacking. Finally, the risks that such complexity and disorganization represented to the utility were inappropriate for the mission critical nature of the utility.

Legacy technology systems (Figure 4.1) had accumulated at AE over the prior 15 years, which posed a dual threat to the enterprise. First, the cumbersome, complex systems caused great frustration for business units that lacked access to the appropriate data to make decisions, but were still expected to support increasing maintenance costs. Second, for a provider of mission-critical

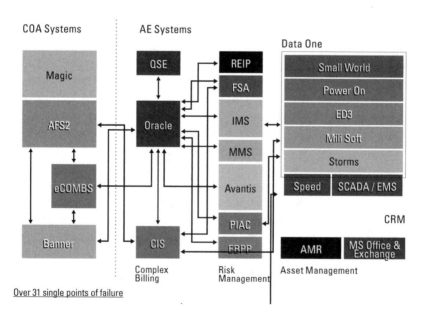

Figure 4.1 AE legacy technology in 2003.

services like a utility, this technology situation created a significant vulnerability in the form of "single points of failure." The wide variety of systems meant that often a single person would have the responsibility and requisite expertise to ensure the operations of a particular application or system. When that person was unavailable, whether it was single sick day or a planned 2-week vacation, the system would also risk becoming unavailable if something were to happen. The analysis documented over 31 mission-critical single points of failure, with potentially fatal disruptions to the utility's daily operations.

Beyond the complex legacy systems and the problems they posed, the analysis highlighted inadequacies in the technology infrastructure that would prevent it from supporting the long-term vision and objectives of the utility. Telecommunications assets in the field and in the corporate offices would need to be upgraded to accommodate advances in digital applications and platforms. Likewise, technology assets in the back office were not set up to manage the massive amounts of data that would be coming their way in the next few years from the addition of a plethora of new data gathering sensors and appliances. Anticipating and preparing for the additional requirements that would be put on the entire organization as a result of massive increases in data became a key driver in preparing for our smart grid journey.

Technology Recommendations, 2003

Based on this thorough analysis, we examined the recurrent themes and with buy-in from other utility executives, created a set of recommendations, which became our road map for 2003 and beyond. The recommendations fell in the following four broad categories:

1. Coordinate
   a. Coordinate the purchasing of technology company-wide.
   b. Coordinate all IT resources to improve service levels company-wide.
2. Simplify
   a. Reduce the number of languages and applications supported company-wide.
   b. Automate key missing processes and integrate with legacy.
   c. Deploy a portal for business managers.
3. Expand/upgrade
   a. Expand network architecture to support e-commerce and any-device access.
   b. Upgrade data centers, security, and disaster recovery plan.

    c. Implement company-wide smart grid architecture.

    d. Build enterprise data bus and data warehouse/data marts.

4. Invest

    a. Invest in quality, documentation, and training.

By early 2004, the picture at Austin Energy was becoming much clearer. In fact, I had gathered sufficient information by that time to produce a technology strategic plan. The value of a strategic plan goes beyond mapping out a vision, to mapping out the frameworks and goals to achieve, and extends to communication of that vision within the organization. The ITT department used the recommendations that came out of the initial assessment and the utility's strategic plan to create five key initiatives for the technology strategic plan: (1) Create and empower a technology governance structure, (2) upgrade and standardize enterprise technology architecture, (3) implement project and resource portfolio management, (4) improve technology alignment company-wide, and (5) increase operational efficiency and quality company-wide.

Accidental Versus Deliberate Smart Grid Architecture Design

Let me diverge for a moment and talk about an important topic to me, one that proved critical to our success at Austin Energy: namely, smart grid architecture, the root of successful transformations. A smart grid architecture, which includes four levels: processes, applications, data, and infrastructure (networking, computers, and data storage), is the critical component to make an enterprise flexible enough to adapt to a dynamic and changing marketplace. At Austin Energy, we were fortunate to have an advance view of the oncoming future, given that we were progressive leaders in a variety of areas (energy trading, green building, energy efficiency, green power, and so forth). We knew that the electricity space was changing into a more dynamic environment, which led us to focus on smart grid architecture as a basis for our transformation.

In my humble opinion, process innovation is the ultimate competitive advantage and also the key to any successful transformation. Consider for a moment two hypothetical utilities, identical from the outset in territory: resources and talent. The only way our two utilities in this hypothetical scenario can differentiate themselves in the marketplace is by the way in which they go about delivering their products and services; in other words, it is processes that are the key differentiator, but beyond the processes themselves, arguably it is their respective approach to process innovation management that actually separates them and determines their different outcomes. The difference lies in how the utilities innovate for process change.

And process innovation is supported by, depends upon, and is enabled by the smart grid architecture. A rigid smart grid architecture begets slow, limited process innovation. Conversely, a flexible smart grid architecture fosters rapid, nearly unlimited process innovation capabilities.

Given that the nature of process innovation is to focus on all transactions to collapse cycle times, improve customer experience, and enhance customer loyalty, it follows that the customer must be an integral part of the transformation effort. Traditionally, that has not been the case among utilities, whose systems have been more concerned with their infrastructure than with the experience of their customers. This focus on infrastructure over customers, by the way, is not limited to utilities; many companies in many other industries focus primarily on applications, data and/or infrastructure architecture to gain incremental improvements. But as we've said, it is investment and attention to process innovation, not just applications, data, or infrastructure investment, that leads to profitable and long-lasting improvements.

The electric grid challenge is to move beyond a traditional focus on internal applications, data and infrastructure—the electric grid, as it were—and to expand the focus to include the customers and their systems, the points of interaction within the systems, and the processes that enable transactional improvements.

Let me give you an example of how this worked at Austin Energy. At AE, we had a single domain, network architecture—a very strong shield that allowed nothing in or out of AE—no Internet access, no remote e-mail, no e-payments, and so forth. Essentially, for security purposes, we had an Intranet sealed off from the rest of the world. We had network architecture that some would describe as "hard and crunchy on the outside, and chewy and soft on the inside." But what may be delicious in a brownie is not so good in network architecture. Four use cases demanded that we adjust our network and security architecture.

The first use case concerned allowing utility employees to access the Internet from within the enterprise to do better research and keep up with the new information services online. The second use case was about the requirement to enable remote email and work file access for the employees when they were outside the office while at home or traveling anywhere. The third use case addressed the need for customers to access their usage information and to pay bills online from home or any place in the world. The fourth and final use case, and the most complex scenario, is to allow for the secure access to customer files by customers who are also employees, while they are at work. The challenge of this last use case is that employees start from a secure environment and use company equipment to go out to the Internet, only to come back in to look at their own energy use information and pay their bills, just as a nonemployee could. But to do that, the network architecture needs to be designed with that particular use in mind. To maintain security and enable all these use cases, we needed to evolve the network architecture to become a multilayer, multiaccess, profile-driven architecture.

We started our evolution by focusing on customer-driven use cases. We had to begin with the customer experience first, which led us to the particular processes that enable those experiences. This wasn't easy, because the cultural attitude inside the utility had been "security trumps everything." We surely recognized the need to maintain security, but we also had to accommodate the demand of new requirements from customers and employees for data access and new services, as we described in the use cases above. Our use case approach led us to processes and requirements and new choices on how to design the applications, data, and infrastructure that would support the processes.

The architecture methodology and design approach that we chose came from the Open Group, a nonprofit organization that specializes in the design of enterprise architectures [1]. Joining the Open Group in 2004 was a critical first step that led to many benefits down the road.

Unfortunately, choosing to design a smart grid architecture was not yet a common practice at the time. In fact, we were breaking new ground as pioneers. We deliberately designed our smart grid architecture based on our use cases and the processes we would need to support. In contrast, the typical path to an enterprise design is totally accidental, even unconscious, as described in Figure 4.2. Starting with the purchase of an enterprise application, the architecture choice gets made by default. Then the second application purchase repeats this process, potentially resulting in a second and totally different default architecture choice. Repeat this process enough times, as is inevitable over months and years, and the foregone conclusion is a complex management challenge like the one I faced when I started at Austin Energy.

The real need for smart grid architecture becomes self-evident when you consider the operational choices a utility makes as well, revealing a link between IT and OT. This diversion is noteworthy because it describes the genesis of the smart grid architecture that we ultimately created. Somewhere in the middle of these events, I had an epiphany. There was an aha moment for me, when I recognized patterns from other industries and other experiences I'd had in the past at companies like Philips Electronics, Digital Equipment, Borland, and Microsoft. My experiences at companies in the telecommunications and computer industries helped me to reach the enhanced understanding of the true importance of smart grid architecture and the connections between IT and OT that I've described in this section.

It was our adoption of the use case approach that led us to recognize that we needed to move beyond IT, and take our lessons learned over to the OT side of the house. We needed to look at operations and segment domains to create a

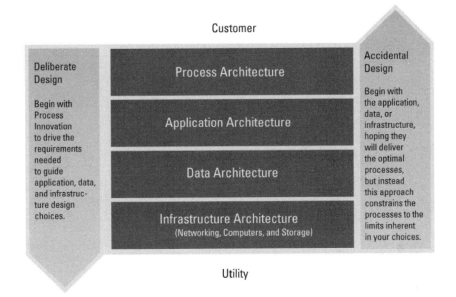

Figure 4.2 Deliberate versus accidental smart grid architecture.

layering of the security architecture. For example, we needed to isolate critical engineering systems such as SCADA/EMS, outage management, distribution management, and so forth and provide special privileges and access. Also, we needed to isolate the billing system, the asset management system, and the financial systems. We went from having a castle with a moat and wall, a relatively simple and secure system—but one that was very limiting—to a multilayer, multilevel, profile-driven access architecture that provided the flexibility we needed to create a smart grid, as shown in Figure 4.3.

Data Flow

A natural follow-on task after crafting our new smart grid architecture was to map the necessary data flows to integrate mission critical utility systems. Such integration is enhanced and optimized by the new smart grid architecture to become a treasure trove of benefits that help make the utility operate more efficiently, increase accountability throughout the utility, and reduce operational and capital expenses by eliminating duplication of effort, manual entry, and paperwork. As stated earlier, the benefits trail starts with a thorough understanding of where the data comes from, how it moves throughout the organization, and what such data flows imply for technology infrastructure planning and management (Figure 4.4).

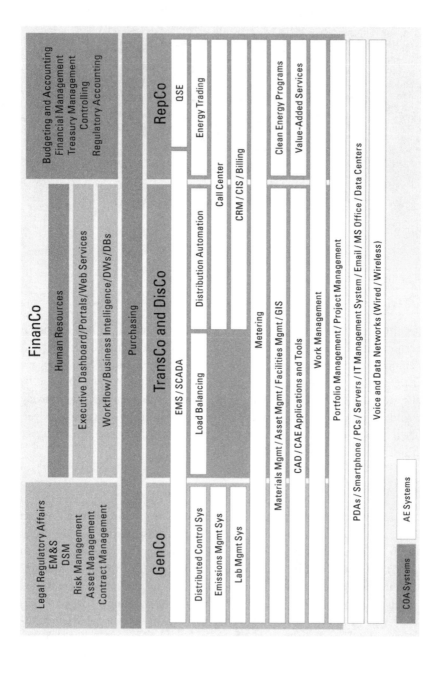

Figure 4.3 AE smart grid architecture in 2010.

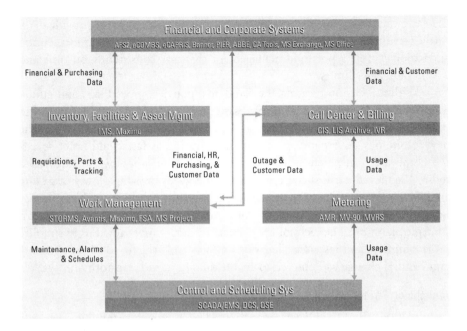

Figure 4.4 AE data flow diagram.

Executive Buy-In and Technology Governance

Beyond the initial assessment to highlight critical problem areas, the early technology realignment process depended on gaining buy-in from all departments. Without universal buy-in and a common perspective throughout the enterprise, we weren't going to go far. To execute on the list of recommendations above, I sought to get everyone on the same page by communicating the technology vision and leveraging inclusive processes. But to begin to instill order out of chaos, it was critical that we create a technology governance plan.

I first created a Technology Leadership Team to ensure good communication with the utility's executive team and provide the necessary oversight mechanism we so sorely needed. And the business unit steering committees we established acted as "best of breed" decision-making panels, so that departmental executives and leaders had greater control over their functional areas, as well as closer communication and control of the technology resources dedicated to them. The creation of an Enterprise Architecture Council, a Technology Security Council, a Disaster Recovery Council, with company-wide representation and managed by IT staff, provided the necessary research, planning, and technology selection options to easily meet business unit requirements. The councils were critical to begin the journey of evolving from technological anarchy to practical standardization, increased productivity, and proactive control. Finally, the early establishment of a project management office (PMO) provided still further structure to allow systematic improvement towards project execution and rational project portfolio management that was consistently on time and within budgetary limits.

With a focus on creating the standards, policies, procedures, and guidelines by which architecture, security, disaster recovery, and data decisions would be made to achieve great efficiency and effectiveness across the enterprise, the new technology governance plan started delivering at faster and faster speeds the corporate transparency, accountability and innovation we needed. In a sense, I achieved early success at Austin Energy by moving slowly, assessing and planning before prescribing and acting, all while gaining valuable information that contributed to ongoing project management as well as longer-term strategic planning. The technology governance structure enabled us to simplify project ranking and resource allocation and was truly the first and most important cultural change we achieved to enable the smart grid transformation at AE.

Technology Strategic Plan

In July 2004, the ITT Department released its first ever technology strategic plan internally. The plan included five key elements: (1) create and empower a technology governance structure (described earlier), (2) upgrade and standardize enterprise technology architecture (smart grid architecture), (3) implement project and resource portfolio management, (4) improve technology alignment company-wide, and (5) increase operational efficiency and quality company-wide. I would summarize our focus back then as shifting from operating a collection of business units to operating as a cohesive enterprise (which is also a critical aspect of advanced smart grids).

The most important technology goal of the plan was to build a smart grid. The steps to build the nation's first smart grid were driven by the need to simplify infrastructure, improve decision making, adapt to faster changing business needs, improve disaster recovery and business continuity planning, improve regulatory compliance, increase quality standards, increase reliability, increase customer satisfaction, and reduce capital and operational costs.

The path to the smart grid ran directly through the technology governance methodology we used to ensure centralization of technology purchasing, decision making, and business alignment, while remaining flexible and driven by business line executives and managers. These leaders acted as project sponsors, accountable to the enterprise for funding of the projects, business cases, ranking and alignment against the corporate strategic goals, and committed to delivering the benefits outlined in the business case that justified the investment.

Three key principles drove the executive support I needed for this transformation. First, we needed to architect enterprise-wide, but deliver one discrete project at a time, in order to demonstrate success, adoption, and culture change and build the necessary momentum for change. Second, we always emphasized that perfection is the enemy of good, where the search for perfect solutions risks forestalling good-enough solutions that would otherwise contribute to progress; we often settled for less than what we hoped for, but acknowledged each time that progress and forward momentum were the keys to sustainable success. Finally, we created a widespread understanding that building the smart grid is a journey and not a destination, a marathon rather than a sprint. This final principle was critical to maintaining morale, overcoming hurdles and keeping the process underway.

The plan emphasized that achieving success would require true top-down commitment to business process innovation and managing and rewarding culture change that optimizes the attainment of higher levels of efficiencies and effectiveness and improved customer experiences. The plan would go on to highlight two key insights, namely that the smart grid can be delivered sooner than most people think, and that availability of technology is not the key constraint to the delivery of a smart grid, rather, any holding back was generally based on managing risk, selecting business models, and addressing internal political issues. The plan also pointed out that the path to success would require a new way of thinking about our challenges as an industry and the solutions needed to empower a total transformation.

Getting the entire organization on the same page—instilling buy-in to the technology vision in addition to keeping the lights on—became the central tasks of 2004. A key element of the technology vision was to leverage common IP networking technologies to promote efficiencies and move the utility away from its traditional orientation around departmental silos, departmental applications, and specialty networks.

We already had installed a fiber backbone, and we had agreed on a strategy to expand fiber out to all of our distribution substations. Also in 2003, we had made the decision to deploy wireless networks to support our AMR and demand response applications. But our wireless networks were not integrated with each other or with the fiber network; and we had not adopted an integrated IP network vision by any stretch. In April 2004, I met John Cooper and began planning a project to promote IP networks internally, as detailed next.

The GENie Project: Considering an Integrated, Shared IP Network

Nothing summed up this new focus on an integrated energy ecosystem better than the GENie Project, which we began planning in May 2004. Over the next year, GENie proved a valuable tool to communicate the potential of a shared wired and wireless IP network and to promote buy-in of the enterprise focus of the technology strategic plan. We chose the name GENie, which stands for "geodesic energy network: information + electricity," because it captured the vision of a foundational IP network supporting new electricity architecture.

I hired John Cooper as the GENie project manager to raise the profile of IP networks inside Austin Energy. John spent the first several weeks assessing the situation, as I had done when I started back in early 2003. He then formed two cross-functional teams. The GENie Project Advisory Council was comprised of departmental management representatives oriented around communications or specific critical applications, providing coordination and guidance in drafting a strategy and implementing a trial. The Network Vendor Advisory Council, comprised of IP network communication and application vendors, served as subject matter experts and joint change agents to help communicate the IP network vision to the utility.

The GENie project highlighted the challenge of migrating disparate AE networks supporting individual applications to common network architecture. The focus on a single, integrated IP network would be a sustaining aspect of the smart grid vision that would emerge over the coming years, but the first step was to support the business case for integrating a wireless IP network, which required a comprehensive review of current applications and networks and ultimately, selecting critical applications for a trial.

As shown in Figure 4.5, an integrated IP network can be the key to enabling a variety of applications, integrating relevant data, and providing significant program impact in a utility environment. The GENie project identified an array of business process improvements (BPI) that would be enabled with a smart grid and set about to demonstrate the hypothesis of BPI and the efficacy

Data + Applications + GENie Network
↓↓↓
Business Process Improvement

App Pilots	More Data	Program Impact: Enhancement + Additions			
AMR	Interval Meter Reads Load Profiles Demand Usage Power Quality	Outage Detection from Manual Reads to AMR		Energy Management Services	
Conservation	Curtailment Status Chiller Flow Data Critical Peak Pricing	Demand Response	Load Shedding	Load as a Resource	TOU Rates
Customer Care	Outage Status Off-Cycle Reads Theft Detection	Outage Notification	Billing Consolidation	Trouble Acct Mgmt	Customer Loyalty Programs
Distribution/ Substation Automation	Outage Status Circuit Status Substation Power Quality	Mobile Data	Reliability Enhancement/ Outage Recovery	T&D Planning	Automated Controls
Security	Video Surveillance	Asset Protection		Homeland Security	

Figure 4.5   Business process improvement and GENie.

of an integrated wired and wireless IP network with a trial. The GENie trial deployed a small Wi-Fi mesh network and ran early 2005 with notable successes and challenges. Looking back at GENie, we realize now that we were ahead of the curve in 2005—much of what the current DOE ARRA projects are doing, we envisioned in GENie, but back then, the technologies we used were less capable and more expensive than they are today.

The GENie project concluded in May 2005 with a report to the ITT department and AE executive team, which included among its many recommendations the deployment of a system-wide wireless IP network that would complement our fiber network and that could be shared among all AE departments. An important lesson learned at this juncture is that timing truly does matter. When the GENie recommendations were made, more specifically, when the principal recommendation to build out a system-wide wireless IP network was made, Austin Energy capital coffers were lighter than usual because of three mild summer seasons in a row (mild summers mean less air conditioning and lower cost utility bills). The recommendation to expand our wired IP network with complementary wireless networking failed to win the approval of AE executive management.

On the other hand, the GENie report did serve its purpose of educating AE senior staff and departmental leadership inside the utility on the benefits of an enterprise-wide focus, continuing a process of paradigm resetting that I had begun with the technology infrastructure rationalization work since I had started at AE. And another key recommendation of the report, to upgrade the one-way AMR system into two-way AMI system, and to expand its coverage to match the entire service territory, proved more appealing and did receive executive approval. The incumbent vendor, Cellnet, now a division of Landis+Gyr, proposed leveraging their underlying 900-MHz network two-way technology by deploying digital smart meters to all residential and commercial customer sites in AE's service territory. The upgrade and expansion from our initial deployment of 130,000 AMR meters began an ongoing series of projects that we would soon come to see as a major plank of our emerging smart grid program.

By mid-2005, the changes I had launched over 2 years before began to open up new opportunities and new levels of service. One of the key recommendations to come from the 2003 preliminary assessment had to do with the lack of visibility for Austin Energy customers on their bills. The ITT department recommended that specialized Web portals be developed to showcase energy information to multidwelling-unit (MDU) managers. During the GENie project, we had spent time assessing the different applications that would be supported by a new enterprise-wide IP network. The potential to expand the partially completed AMR system, which at the time was deployed only to MDUs (i.e., apartments and condos) throughout the service territory, stood out.

It wasn't just about AMI though. The requirements of an expanding family of applications kept us busy planning and building strategy. Mobile workforce applications such as mobile laptops would benefit by pushing access to complex system data out to the edge, enabling field workers with new potential for enhanced productivity. Demand response also held great potential, given the significant consumer adoption and widespread success of the free thermostat program at that time, which communicated via a digital paging network. However, while the GENie project had been successful in highlighting the potential of an enterprise-wide IP network, the departmental approach of single purpose networks to support silo applications still had significant momentum and the vision of a unified, integrated wired and wireless IP network remained that—an aspirational vision.

Project Management

The work of the ITT department is conducted through programs and projects, making the establishment and institutionalization of an effective project management office (PMO) a critical step on the way to a rationalized technology infrastructure. Without effective project management, it becomes routine for deadlines to slip, budgets to pass limits, and scope creep to increase the ongoing project requirements. Project tasks slip past their deadlines and completion dates disappear over the horizon. We needed a PMO, and fast.

When I started, we had two people on the ITT staff certified by the Project Management Institute (PMI) [2]. PMI certification is widely regarded as a critical measure of quality, so one of the first things we did was send staff members to the PMI to get certification (today the utility has over 50 employees who are PMI certified). We followed PMI standards and created an online PMO that allowed us to create easily accessible templates, reports, and dashboards, representing such project metrics as "customer usage" and "projects completed," as well as monthly steering committee reports.

In 2009, of over 4,000 companies analyzed by Gartner regarding project management maturity, only a handful worldwide had achieved Level 4 maturity. I'm proud that AE's ITT PMO was among that elite group.

In PMI process flow, as detailed below, there are five major states or "gates" whereby projects can be tracked: selection, initiation, planning, execution, and closing, with additional gates within each of those steps. Project progress is tracked in a PMO using the project state, its position on the project schedule, and the project complexity. Project criteria qualify projects as either "Run" (i.e., a project about keeping the light on), "Grow" (i.e., a significant enhancement to an existing solution), or "Transform" (i.e., a project to replace an existing solution with a new one).

The PMO allowed us to go from not tracking hours or effort at all when I arrived, to tracking every hour and every dollar, matching those to every requirement for every customer and every project to deliver the defined benefits. When I arrived, we had a ratio of 90% Run and Grow projects to keep the lights on, and only 10% Transform. By 2009, we had moved that ratio to 60% Run/Grow and 40% Transform. In other words, our standardization, process, control, and best-in-class practices had freed up an additional 30% of resources to apply to Transform projects.

Service-Oriented Architecture

In 2005, the ITT department also took the first steps on another significant project, the transition to a service-oriented architecture (SOA). An SOA approach provides innumerable benefits over the long term for an IT department and the organizations it supports. I decided to power Austin Energy's smart grid using an SOA that would follow the principle of delivering presentation, process, and information as services to all stakeholders (including central power plants, distributed energy plants, the wholesale energy system, the transmission and distribution grid, the meters, smart appliances at customer sites, electric transportation, and the delivery of timely information via portals to all customer types).

In May 2006, we went live with the SOA-generated desktop application for call center representatives, integrating the billing system with the outage management system. By allowing a function to appear once but be usable for any application needing it, the SOA eliminated extra steps in call processing, such as ensuring the completion of a customer validation within the outage management system. Such validation is accomplished within the billing system, the work management system, and the financial system. Historically, network architectures would provide that every application must have all these services wrapped within the application. But in a service-oriented world, the first step is to create one customer verification service, then make that service available to any application that needs the service.

For example, consider these steps in the outage management system (OMS). First, a customer calls to report a power outage, and then the application verifies the customer's location and extracts the customer's status from a database. As soon as the call center agent transfers the information to a work order, the information travels to the OMS, which dispatches a service truck. Where this entire process originally took about five minutes to complete, with repetitive steps over multiple systems, the new approach took that time down to 1.5 minutes on average, a reduction of about 70%. The process took five minutes because an employee would need to take such manual steps as looking up different items in different databases and applications, before he/she could eventually push a button to proceed to the next step. The cost savings were significant, but the improvement in customer service experience was even more dramatic. Before the SOA implementation, the application could handle only 4,000 calls per hour and required considerable waiting time on line for customers. With SOA in place, on the other hand, the application could handle as many as 50,000 calls per hour. Similar if less dramatic efficiencies were also found in the back office.

Where traditional applications required programmers to install many of the same functions for each process in an application, using an SOA shrinks the overall amount of code used, standardizes functionality, and minimizes mistakes as employees use the same data sets.

The service-oriented architecture effort inevitably drives improvements in a network, computer systems, data storage, data schema, and business process architecture layers. Doing one system and delaying others costs more in the end, and takes longer. The SOA design process starts by mapping the current leading business processes, which provides valuable insights on the status of the enterprise. A session to set ambitious stretch goals follows. Gaps are identified and a map is devised to proceed from the current to the future state, with quick win projects identified and given priority. I found it important to stay focused and work to the plan, and document lessons learned along the way. Finally, to maintain morale and generate momentum, we celebrated wins regularly and highlighted milestones throughout the organization as they were achieved, so that the entire organization understood the progress and could participate in the transition.

These three steps—concluding the GENie project, adopting a system-wide AMI network, and installing SOA architecture—provided the required foundation for the future deployment of a smart grid at Austin Energy. The GENie project helped to instill a shared vision among utility leadership regarding a more holistic enterprise approach and the potential to leverage IP networking to gain efficiencies in utility applications; the AMI network provided an opportunity to take a bold first step on that journey, and the SOA architecture provided sustaining benefits of long-term cost savings and enhanced departmental functionality.

Standards and Quality

Standards and quality are vital components when your goal is to become "the utility of the future." This vision, pursued by the ITT department as well as the entire organization, was stimulated and accelerated by our devotion to standards and quality.

The International Standards Organization (ISO) provides a template for organizations with its ISO family of standards on quality management, which are designed to help organizations ensure that they meet the needs of internal and external stakeholders. From the outset, we established ISO certification as an organization goal, a unique objective in the utility industry. By 2007, we had reached one of several goals, becoming the first electric utility in the United

States to achieve ISO 9000 certification [3]. This was an important milestone for many reasons, but one reason stands out in particular. Achieving this recognition became one in a series of external validations of our progress and success in our transformation. External validation proved critical to change the organizational culture and generate interest towards continuous improvement and sustained momentum.

Similar to ISO, the Information Technology Infrastructure Library (ITIL) is a set of concepts and practices for IT professionals to manage and run their operations better, providing detailed descriptions of IT practices and comprehensive checklists, tasks, and procedures that can be tailored to meet the specific needs of an organization. We chose to use ITIL to promote a deliberate transition to a more organized and effective department. Service support and delivery, infrastructure management, security management, business management—all these capabilities and more benefited from our use of ITIL. Each year, we hired external consultants to evaluate our progress on the ITIL continuum, measuring how we had matured as an IT organization according to the ITIL maturity model. Again, such benchmarking and external validation proved to be key tools to motivate continued progress among the ITT staff, but also to bear witness of our progress to the other departments inside AE.

The On-Line Service Catalogue is a great example of a tool that was inspired by ITIL; we published the catalog for our internal customers, starting with a hard copy version issued in September. Soon thereafter we were able to showcase with greater ease online the different service categories the ITT department provided within the utility. The catalog provided an overview of departmental mission and vision, an organizational chart, and also showed how the ITT department interfaced with other departments through the variety of governance mechanisms described above. The online catalog helped to move the utility along in our transformation to a paperless, computer-based organization, described in more detail in the creation of our digital platforms next.

Digital Platforms and Data Access

At the beginning of 2006, I became more focused on making the applications we had been creating more user friendly and effective as tools that would promote better management practices, such as management by objectives (MBO) using real-time performance measurements, and feedback. For instance, we launched a project at that time that was focused on building user dashboards to leverage ROI tools, an internal project I pitched to leverage all the data that would be coming in from the different devices and applications we were deploying.

At the time, much of the data reporting was accomplished using paper reports populated with traditional data tables using month-old data, which required significant effort to produce, analyze, and interpret. To improve on executive data presentation, we would need to better understand data flows and formulate a data strategy. Tracking performance to objectives required the creation of key performance indicators (KPIs) that would provide the data hooks for user dashboards. We were able to develop dashboards that became standard for managers throughout the utility, providing access to critical performance data in a more accessible format on a far timelier basis.

Snapshot 2007

While the ITT department continued to promote a cohesive ecosystem approach based on the technology architecture vision, the cultural adherence to organizational silos remained a challenge. The inertia behind such organizational segmentation in an electric utility should not be underestimated. Remember, while ITT recommended a universal wired and wireless IP network to overlay the grid at the conclusion of the GENie project, capital constraints and executive sentiment directed the program more along the lines of incremental additions of applications like AMI expansion, with distinct ROI objectives and project plans that conformed more closely with the business plans of the silos.

By the end of 2006, the ITT department saw the year's highlights divided in three major categories, all related to making more data available for better decision making and lower costs. First, we upgraded technology infrastructure continuously, from the network out in the field to the back office. Second, we shifted more and more to digital platforms, with emphasis on Web portals for remote access. Finally, we leveraged network technologies and infrastructure to provide greater remote access for field operations. In this way, AE's smart grid began to emerge through a variety of incremental improvements, all oriented around leveraging digital technologies and improved network access to achieve AE's explicit mission to become the utility of the future.

The Smart Grid Emerges as a Tangible, Explicit Utility Theme: 2007–2008

By 2007, we were completing many of our initiatives and while we had made considerable progress, we still had a long way to go. Roger Duncan's vision of a utility more reliant on distributed energy resources became more prominent with continued program success, particularly with regard to the nascent campaign to promote plug-in hybrid electric vehicles or PHEVs (what we now call EVs). Smart thermostats had proliferated to such a degree that the utility had 100 MW of curtailable load when needed. In keeping with the digitization of our processes, we focused on completing the integration of an electronic bill payment and presentment system in 2007.

Smart Grid Infrastructure and End Device Integration

The emerging smart grid at AE in 2007 still resembled a collection of programs, processes, systems, applications, and sensor devices, more than a truly integrated ecosystem. However, if you looked closely even back then, you could see an interconnected smart grid ecosystem emerging, consisting of the mapping of processes, data, schemas, field applications, head end equipment, data center equipment, back office applications, and networks. All this activity was supported and orchestrated according to smart grid architecture we had laid out years before, so critical to our success.

First, a backbone of fiber assets connected all substations by 2006, with subsequent incremental additions. An assortment of wireless networks covered much of the service territory supporting various applications. In 2007, we also ran a broadband over power line (BPL) pilot, which confirmed what we expected—BPL worked but it was too expensive. In 2006, we had also deployed a small downtown MetroMesh Wi-Fi project to support the World Congress of IT (an international conference Austin hosted that year). The Wi-Fi mesh project remained as a city asset in downtown Austin, which AE supported.

Beyond networks, the emerging smart grid in Austin was comprised of head end equipment, principally AMI smart meters (over 410,000 when fully deployed) and smart thermostats (ultimately over 100,000), but also including a small but growing number of smart devices attached to substation and distribution automation gear. We also began planning in earnest to meet Roger Duncan's vision, anticipating the integration of connected electric vehicles, energy storage and solar PV systems, as well as home energy management system (HEMS) devices that would emerge over the coming years in the smart home. In 2006, we had launched our Web site to promote the plug-in hybrid electric campaign nationwide and interest continued to grow in PHEVs, with retrofit kits brought in to help us start learning more about PHEVs. We also upgraded the Green Building Web site and launched an e-newsletter. Finally, in 2006, we had taken over the energy planning of the city's traffic lights, continuing in 2007 with incremental replacements of incandescent lightbulbs by LEDs connected to an intelligent control system.

Completing the use of IT in the field were field mobile data applications, such as mobile mapping software and devices like laptops and later, smart phones that required connectivity and support for an emerging mobile workforce. A key challenge to providing data to field-based employees was the large bandwidth required for maps in service orders and trouble tickets. Consequently, initial applications required the workers to load high-bandwidth data on their devices before heading out to the field for the day.

Moving from the field to the back office, our SOA deployment continued and we added asset management software, notably the Maximo software implementation that year. In 2007, we began a significant effort to build and integrate a data warehouse business intelligence program and related data centers. In our corporate facilities, we replaced our old PBX telecom system with new technology by Avaya, complemented by Cisco technology for our transition to voice over IP (VoIP) inside all AE facilities.

Beyond the ITIL compliance discussed earlier, we continued with business process improvement, change management, and communication initiatives. As a noteworthy aside, we had good success gaining organization commitment through the use of operational level agreements (OLAs), similar to service level agreements (SLAs) that vendors commonly use with their customers to define the service they provide, but in this case, ours were internal agreements within the different technology groups. We upgraded our customer portals for MDU customers and in our external communications, we began to articulate the utility goals of carbon reduction and clean energy, highlighting our utility of the future vision, where electricity demand and supply would be equally managed and accounted for.

Snapshot 2008

By 2008, after I had been at AE for 5 years, we could look back on significant success. The start of that year marked a landmark that we saw as the completion of our "4-year transition." The 2008 ITT Strategic Plan, which forecast activity from 2008 to 2011, included a graphic of "Austin's smart grid," which was aspirational at that time—we hadn't yet completed all our work, but by then, according to our organizational vision, we were indeed building a smart grid energy ecosystem. The network portion had been completed by 2008, and the infrastructure and key applications to enable the smart grid had been procured and deployed (data centers, major systems and applications, and so forth). The missing core element to complete our smart grid was integration, which would be the bulk of our work going forward.

By 2008, we had established a strong brand for Austin Energy with the utility of the future vision and our considerable progress to deploy a smart grid. As a city-owned utility, we not only were intent on leveraging the best practices and best technologies available, but we also felt a responsibility to stay involved with a variety of leadership groups nationally. Back then, I was a frequent speaker nationally (and occasionally, internationally) describing our progress at AE. Among the groups to which we contributed regularly and that we used to stay abreast of developments were a Texas-based group, the Center for Commercialization of Electric Technologies (CCET) [4]; as well as such national groups as the Large Public Power Companies (LPPC) CIO Task Force [5], IDC's Energy Executive Council [6], the Utilities Telecom Council [7], the Grid Wise Alliance [8], the National Institute of Standards and Technology (NIST) [9], the Department of Energy (DOE) [10], and the Federal Communications Commission (FCC) [11]. We contributed to the discussion where we could and gathered immeasurable help from our colleagues in these and other organizations, highlighting another key point: we are truly on this journey together, and we all benefit from sharing where we can.

As with ISO and ITIL, involvement with these organizations and the seven national technology awards we won over these 4 years (i.e., *Computerworld* Top 12 Green IT Company, *InformationWeek* 500, *HITEC* Top 100, *CIO Magazine* CIO 100 Award, *Computerworld* Premier 100 IT Leader, *Computerworld* Best in Class of Premier 100, and the Association of Information Technology Professionals IT Executive of the Year) provided the vital external validation that proved so important for us when conducting such a long-term program of complex megaprojects.

During our telecommunications and SOA transformation, our short-term goals in the ITT department evolved and to accommodate our progress, we had to recast them as we went along. But our mission remained steady: To deliver the first fully-integrated and self-healing electric utility in the United States. The progressive approach we took with the technology infrastructure and the recognition we began to receive at this point were on par with the accolades other AE departments were receiving at that time for their own progressive achievements in energy conservation and incorporation of green energy into the resource portfolio.

In 5 years, we had transformed ITT from a challenge to a strength, and in so doing we prepared the utility to leverage the investments we had made in applications and systems to improve substation automation, distribution automation, metering automation, energy scheduling and trading, customer service, decision support, vendor management, and mobile field service, among other achievements.

Since the start of the decade, most of our applications and systems had needed fundamental upgrades or changes to keep up with technological advances and the growing requirements placed upon the ITT department. But while we continued to upgrade key applications and systems, we had completed much of our foundational work by 2008 and could shift our focus to fulfilling our mission of building an end-to-end smart grid (extending from the central power plant and distribution system in-front of the meter to end consumer devices and distributed energy resources located behind the meter).

And it's not like we were trying to hit a stationary target either—we would also need to achieve compliance with new FERC and NERC regulations on physical and cyber security and meet ERCOT's redesign of the Texas' wholesale electric market (a transition from a zonal to a nodal market) [12] as well as upgrade to a new digital billing system, expand our mobile workforce management and dispatch capabilities, and so on.

By 2008, we could see that our transformation had allowed us to consistently eliminate complexity and offer new services, increased our service quality to both internal and external customers, increased transparency and accountability of both technology and business lines, eliminated all those single points of failure we had identified back in 2003, replaced legacy applications with state-of-the-art new technology, and increased trust throughout the ecosystem, as evident from the year-over-year increase in customer and employee satisfaction scores that both AE and the ITT department had achieved in the previous 4 years in annual surveys.

Having achieved much of what we set out to accomplish, the AE executive team created a new set of recommendations for 2008 going forward, which I've grouped into two major categories:

1. Adapt to External Environment
   a. Drive and support ERCOT nodal market enablement.
   b. Implement enterprise cyber security based on FERC/NERC and industry standards and best practices.
2. Implement, Simplify, Expand
   a. Implement company-wide smart grid architecture.
   b. Simplify and standardize applications into an N-tier framework.
   c. Expand wireless network services.
   d. Deliver business intelligence platform/framework.
   e. Implement BPI (RUP, SDL, collaboration tools, and so forth).
   f. Deploy B2E, B2C, and B2B portals.
   g. Improve resource transparency and management.
   h. Mature ITT's quality initiatives (ISO/ITIL/CMMI).

New Goals

These recommendations were closely tied to our larger, overarching goals for building a smart grid at AE. Most people now view a smart grid as a technology project, when in fact the smart grid is a means to a much larger end. The smart grid enables a utility to achieve its larger strategic goals, which for AE included this list, outlined below. At Austin Energy, the smart grid was understood as a destination made of multiple projects to achieve our corporate strategic goals.

First, financial integrity tops the list, including overall reduction of capital and operating expenses, which could be rephrased as, "being good stewards of the investments made by the customers, who pay for the utility in rates" and for investor owned utilities, "being good stewards of shareholder investments." Our 2010 financial goal was to achieve AA bond rating.

Second, customer excellence, which includes engaging customers in a new relationship with the energy they consume, and increasingly, the energy they produce, leading them from being passive and ignorant about electricity, to growing active and aware, and finally to becoming responsive and committed partners in energy consumption and production. Our 2010 customer satisfaction goal was to achieve 83% on the Customer Satisfaction Index of the JD Power Ratings for Utilities.

Third, reliability excellence is captured in such utility metrics as SAIDI, SAIFI, and other indices. Our 2010 reliability goal was to achieve 40 minutes for SAIDI, and 0.5 interruptions for SAIFI. (From 2003 to 2008 AE managed to achieve the lowest SAIDI in the nation, which was very gratifying for all of us.)

Fourth, significantly reduce the carbon footprint, as the production and distribution of electricity are principal contributors to CO_2 emissions worldwide. Our carbon reduction goal was captured by a goal to become carbon-neutral in our fleet by 2012, but also a 2020 goal to attain 800 MW of energy efficiency (26% of our total generation capacity, which would allow us to avoid building more power plants).

Finally, integration of renewable energy, which accepts that the smart grid is an enabler of the transformation of the grid from a system based on burning fossil fuels to one that produces power based on technology and natural energy sources. Our renewable energy goal for 2020 was to reach a Renewable Portfolio Standard of 35%, including 200 MW of solar energy generation (6% of total generation capacity).

After 5 years on the job and 4 years in a serious transformation program, working with all my colleagues I had accomplished the implementation of foundational projects and rationalization of processes according to our shared goals, and the ITT infrastructure began to more closely resemble a cohesive ecosystem than a collection of systems and processes. With a sound foundation based on SOA and technology governance mechanisms, the ITT department could truly begin to provide the rest of the utility, specifically the OT managers, the required leadership and the data access and management tools they needed to reach our organization goals and realize our utility of the future vision.

Steps to Integrate an Energy Ecosystem

To achieve the vision of a smart grid, or more specifically our goals and our vision at AE, it became critical to integrate the various components and systems of the smart grid architecture so that the energy ecosystem would operate as a unified mechanism, not unlike the way that a variety of high-quality musical instruments in the hands of accomplished musicians make beautiful music in a symphony, according to the notes in the pages of the music and the guidance of the conductor's wand. Our work in building our own orchestra was laid out in our 2008 strategic plan, which featured the following five key strategic initiatives. These initiatives serve as our guide for this section, where I describe the final steps to integrate our energy ecosystem. They constitute a planning tool for anyone who would seek to do as we did—integrate technology systems to make a smart grid.

1. Create and Empower Technology Governance Structure

"Technology should not drive projects; technology is not in the driver's seat of this bus." This was the credo driving our technology governance structure. Going back to the smart grid architecture section earlier in this chapter, remember that the design of smart grid architecture starts top-down with the customer use cases and moves on to the processes needed to achieve those objectives, then on to the applications, data, and infrastructure needed to support those processes. In contrast, execution of the smart grid design starts bottom-up, with the infrastructure, moving on to data and applications, then to the processes needed to meet the desired objectives regarding customer impacts.

Starting with customer impacts was key to achieving the customer satisfaction goals described above, which drove the creation of customer use cases and objectives including this partial list: (1) managing outage restoration and notification times, (2) receiving usage information to better understand and manage customer bills, (3) being able to participate in energy efficiency and demand response programs, (4) improving timeliness and accuracy of bills, (5) promoting turn-on and turn-off services, (6) enabling customers to manage smart home appliances via the Web or a separate display, (7) being able to participate in variable pricing programs, and (8) selling excess energy back to the grid.

In 2008, we went back to the technology governance plan we had instituted back in 2003, revising it to include an Enterprise Data Council, which defined and managed the policies, procedures, standards, and guidelines for the creation, use, and management of data across the enterprise. As they had since 2003, the different groups within the technology governance plan proved instrumental in maintaining our focus on our objectives and coordinating activities within the enterprise.

2. Upgrade and Standardize Smart Grid Architecture

As I've detailed before, the smart grid architecture plan was the critical piece to drive our smart grid transformation. The biggest thing to emphasize here is that we were rebuilding an airplane while it was in the air—no small feat when you think about it. This is a principal challenge to utilities around the world: to transform their fundamental architecture even as they use their system to maintain reliability and keep the lights on at every moment. To accomplish this difficult task, we first created a parallel track.

Another way to look at our challenging task is to consider what highway departments go through when they build a new highway. We had to build a series of new superhighways right next to old highways, but with more lanes and far greater capabilities. Our focus was on redesigning the network, servers, data storage, processes, and controls, while upgrading applications and facilities. Successful transition required significant planning in advance, a large number of system integration projects, and a balancing of transition with cost management; we always to keep financial impacts in mind, to stay within budgets. We designed an upgrade path that went system-by-system, where we created a new system, then transitioned operations to the new system, taking time to test and confirm the transition before moving on to the next system and repeating the process. In this way, step by step, we evolved our smart grid architecture to a new, more modern and efficient design, while maintaining reliability and continuity of operations.

3. Implement Project and Resource Portfolio Management

The key to running a successful organization, and more specifically, transforming an organization, is to communicate the vision, then translate that vision into achievable steps and coordinate the activities of a variety of different organizations and individuals with widely divergent perspectives, according to a rational timetable and set of detailed instructions, while managing to a budget. We initiated such a rational approach back in 2003, and we began to enjoy the fruit of our early efforts later, as we saw the organization begin to transform along the lines of rational planning and execution.

We created processes according to guidelines from the PMI, and implemented them with the help of our online tool, ITT PMO Live. The Web site provided real-time tracking of the myriad projects we were running, which helped us to manage our processes, projects, and documentation efficiently. A key tool in ITT PMO Live was the project management office workflow, which became our engine to evolve ITT functions and keep the rest of the organization informed and in tune with the changes we were implementing.

Projects followed six steps: select, initiate, plan, execute, control, and close, which took a project from beginning to end. It's important to note that these steps focused on the relationship between the line of businesses (sponsors and customers) and the technology teams. Back and forth, changes were implemented and the tool was updated. We needed to track such things as project definition, requirements gathering, schedule, resources, alignment, justification, prioritization, and funding—for all of our many projects. The benefit of an online tool like this may seem self-evident, but let me emphasize anyway that it was this transparency and communication of the details that gave the organization the confidence it needed to go through with this transformation and ensure that all the day-to-day operations experienced minimum to no impact from the changes we were making. A key output of this tool was the summary report that allowed the entire organization to enjoy a shared understanding of the transformation as it was taking place.

Another key aspect of managing the transformation of a complex organization like AE was managing to a budget while achieving a large set of objectives, balancing business objectives, corporate goals, and technology realities and constraints. Portfolio management became the tool by which we managed to priorities, balancing processes, practices, and specific activities to perform continuous and consistent evaluation, prioritization, budgeting, and final selection of investments. This approach allowed us to provide the greatest value and contribution to help AE achieve its strategic objectives while balancing sometimes competing organizational interests, issues, politics, and agendas. We were able to make trade-offs among competing investment opportunities based on rational evaluation of benefits, costs, and risks.

4. Improve Technology Alignment Company-Wide

Alignment. Let's pause for a second and consider what an important concept that is to an organizational transformation. When a school full of students needs to evacuate the building rapidly in a crisis, they get in lines, exit the building, assemble on the schoolyard, and count heads. In a fire drill, everyone has a role to play and frequent rehearsals ensure that alignment will be second nature if a crisis ever occurs. A fire drill is a good way to explain the alignment methodology we took at AE to accomplish our objectives, but we weren't rehearsing for an event that might never occur. We underwent frequent fire drills, where we reviewed and accomplished successive steps to stay in alignment as we transformed the organization into a utility of the future via the implementation of a smart grid.

We looked at our transformation plans and recognized that we had two levels of alignment in order to implement the necessary technology solutions at AE, starting with our corporate strategic goals. First, alignment at the company-wide level was needed to optimize how we used our finite resources and minimized costs along the way—we had to remain efficient as we spent the public funds that kept us going. Second, we needed to align at the business unit level to maintain operational efficiency, human productivity, and meet customer needs: we needed to keep our business running as we transformed it into a new type of organization.

First, we needed to align the corporate goals with team and individual efforts. To achieve that, we chose what is called a waterfall method. Like a real-world waterfall, we envisioned repetitive processes that flowed downhill on paper based on successive completion of a series of tasks, just like a real waterfall flows down a cliff, pushed along by gravity. We captured team objectives and goals, and action plans for working groups and individuals into planning charts that guided our activities. Detailed planning allowed us to marry individual accountability in tasks, actions, and goals to ensure collaboration with other individual tasks, actions and goals to achieve our organizational goals.

Second, we needed to align business unit goals and strategy with technology resources and plans. So we chose an agile method. As we aligned our technology objectives with business unit goals, we leveraged the tools we had created earlier, such as our principal tool for collaboration, the ITT/BU Steering Committee. In those meetings, the business unit leaders met regularly with ITT staff, which helped them make technology decisions in alignment with both their business unit strategy and vision and the smart grid architecture and our principal goals of "cheaper, faster, and/or better—you can only pick two." In this way, ITT became a trusted advisor as we replaced legacy systems with new technologies, in a virtuous circle of short-term and long-term successful collaboration.

Our agile method had at its core a step-by-step infinite loop with conditions were we systematically obsoleted old products and systems, created new projects, and mapped them to our architecture, operations, and customer requirements, always aligned to our goals. If we hadn't taken the time, if we hadn't dedicated ourselves to spending hours and days meeting to define how we needed to do what we did to meet our objectives, we would have wound up following the multiple processes and methodologies unique to the different departments. We would have run into irresolvable conflicts that would have slowed our progress and stymied our efforts to reach our shared objectives.

5. Increase Operational Efficiency and Quality Company-Wide

This section touches on one of the biggest challenges of implementing a smart grid vision. How can an organization balance the daily needs of running an operation with the vision of executing a strategy to transform the organization, all while staying within the boundaries and constraints of limited budgets? Today, many rate cases around the country contemplate raising rates to finance their smart grid plans. We didn't have that luxury at AE at the time, although we should have. Nevertheless and given our strategic goals, we crafted a plan, as described in this section, to execute an evolutionary path to upgrade our infrastructure, data, applications, and processes according to a smart grid architecture, on a pay-as-you-go basis. Much of our financing necessarily came from savings from new efficiencies and optimization. We did devote new budgets to our upgrades, but they were relatively small compared to some numbers I see in rate cases around the country today. We looked for creative ways to shift budgets to the new paradigm, thereby freeing up capital to finance new projects and keep the transformation on schedule with current and reasonable new budgets. For example, recognizing that we were spending around $4 million each year for new meters to accommodate system growth and address meter replacements, we began to buy the new meter technology instead, even though we were out ahead of our AMI transition and network deployment.

I challenged the ITT department with the goal of running IT like a business. I started speaking on this topic in 2004, giving innumerable speeches where I described the process of taking our ITT department to ever higher levels of maturity in its evolution model, with the last stop being that mantra, "running IT like a business." In this new paradigm, the organization maintains a clear focus on controlling its resources and customers, measuring outcomes and learning from its mistakes and successes. "Lessons learned" became a standard feature in our meetings, because we were educating ourselves at each step of our journey. We encouraged transparency in the delivery of services internally, in our change management execution, and in our optimization efforts, where we strove to achieve better yields at all times. We looked for ways to reduce our operating costs and studied what worked and what didn't, always stressing the importance of being 100% accountable. In this way, we worked to make the ITT department an adaptive, fully optimized organization.

Specific steps we took to achieve our strategic objectives while staying within tight budgets were guided by the concept of balancing supply and demand. To accomplish this objective, we would need to get a handle on the demand for technology services through methodical, systematic control mechanisms with a finite supply of technology personnel and limited time. We did this by focusing on achieving operational efficiencies and improving quality practices. This took discipline, teamwork, and dedication. Given the dynamic nature of the environment and the shifting road map of a transforming organization, our ITT business unit was going to fluctuate with the needs of the multiple projects we ran over these transformative years.

With the concept of balancing supply and demand in mind, we restructured our organization to better manage relationships and resources. We needed to deftly manage expectations of business units that needed new technology resources to keep up with the increasing demands of their business lines; we did this by creating the new positions of relationship manager and resource manager.

One focused on listening to internal clients and building trust, becoming business line champions and their voice inside the ITT department, focused on outcomes from the project managers and business analysts. The other defined the boundaries of what was possible at any given time based on our finite resources, rationing the resource inputs to stay within well-defined budgets.

Over time, we formalized this approach with the PMO and a new customer relationship management organization. Along the way, we drew inspiration and guidance from ISO and ITIL methodologies and best practices. To focus on just one area, software development, we followed the Capability Maturity Model developed by the Software Engineering Institute at Carnegie Mellon to assure a sophisticated and efficient software development life cycle. We adopted the Rational Unified Process (RUP), a process standard that is widely recognized around the world.

The year 2009 saw the realization of many strategies as our smart grid came together. Inevitably, the world started to catch on with what was happening at Austin Energy and our smart grid journey to build the utility of the future. Our success had spilled over and we were getting more awards that one could have expected. We received 11 awards in 2009 recognizing our smart grid leadership and contributions to mankind (e.g., UtiliQ #2 Company by *Intelligent Utility* magazine and IDC, *Computerworld* Top 12 Green IT Company, Energy Central CIO of the Year, *CIO Magazine* Hall of Fame Finalist, *Computerworld* Honors 21st Century Achievement Award, *Computerworld* Honors Finalist, *Computerworld* Honors Laureate, *InformationWeek* 500, UTC Apex Award Finalist, HITEC Top 100, and *Hispanic Business Magazine* Top 100 Most Influential).

As I close my reflections on our progress, I can't help but recall the challenges AE's GM Juan Garza outlined when I took this job back in 2003. He described an organization that lacked the funds to make the changes it needed to make. We found those funds by creatively changing the basic assumptions we were operating with, following an old adage I am fond of: "When faced with an intractable problem with no good solutions, revisit your assumptions. Sometimes changing the problem itself leads to new solutions." In our case, we addressed the way the organization operated. We changed our approaches to be more inclusive, transparent, and accountable, and used such tools as creativity, standards, best practices, and external validation to drive the behavior changes that would release the funds we needed to help finance our transformation.

Lessons Learned

The case study presented in this chapter reveals a pathway to build a smart grid. Given the pioneering nature of this case study, we characterize this smart grid approach as "1.0," recognizing that it offers a number of lessons learned, with two highlighted next.

Smart grid architecture design is a necessary first step.

A principal lesson learned is that approaching the creation of a smart grid by incremental additions of applications is difficult, complex, and expensive. Stand-alone application choices with dedicated tools and networking resources cheat the utility from its future by failing to devise a complete blueprint for a smart grid architecture, which would follow these steps: (1) design the smart grid architecture first, (2) define the necessary use cases, and (3) review all the processes, selecting the applications needed to achieve the blueprint. That said, you have to start somewhere, and there were many debates along this journey

concerning which system was at the core of the smart grid architecture. Choices
were the geospatial information system (GIS), the asset management system,
the workforce management system, or the utility's control systems (SCADA/
EMS or DMS). After many what-if scenarios and debates, managers agreed that
there was no one single way. In the end, the managers at AE chose the geospatial
information system as the core.

Leveraging public networks has appeal if necessary conditions are met.

Another important learning is that utilities could leverage public networks to
achieve their smart grid. At Austin Energy we met many times with public
carriers to explore a partnership. They only needed to provide us with four
things to dissuade us from building our own networks. These four elements
are: (1) Cost (fee for monthly access per device), (2) coverage (need to provide
access to every required device throughout the service territory and with multiple networking technologies an option), (3) adequate service level agreements
(for priority access and restoration of service), and (4) commitment to deliver
network access to every end point for a minimum of 10 years. These four elements proved difficult for public carriers to meet from 2003 until 2009. As this
chapter is written, several public carriers across the globe appear to be ready to
meet the above requirements.

The vision of an advanced smart grid as a new, improved approach to
building a smart grid flowed directly from the experiences at Austin Energy
from 2003 to 2010. The transition was complicated by the mistakes made and
diversions taken along the way. For instance, we only recognized after the fact
that our adherence to an applications-first methodology was a principal culprit.
"Hindsight is 20/20," as they say, and for us, hindsight revealed multiple insights. The principal insights we gained could be summarized thus:

> The considerable advantages in investing time and resources to get the
> smart grid architecture in place, to get the organization behind the
> transformation effort, and to design and build (or lease) a foundational,
> integrated wired, and wireless IP network were keys to our success, by
> enabling much simpler and cost effective deployment and integration of
> applications.

Highlights on the AE Smart Grid Journey

To recap, the following list of highlights represents the key lessons learned and
achievements of the smart grid program at Austin Energy.

First, the smart grid architecture became the primary organizing principle
for Austin Energy's smart grid transformation, which guided AE to deploy such

components as portals, enterprise service bus, data warehouse, business intelligence, cyber security, project management tools, fiber backbone, and so forth.

Second, the upgrade of existing one-way wireless networks in 2007, and expansion of coverage to the entire service territory, enabled full connectivity to every device in the field, starting with the AMI system and its 410,000 smart meters.

Third, the deployment of nodal market tools to accommodate the changes at the ERCOT wholesale market included a new generation management system (GMS), a new network modeling system, an upgrade to the SCADA/EMS system, and, a new asset management, inventory management, and material management system.

Fourth, the deployment of smart meters throughout the service territory, reaching 100% coverage by the end of 2009, brought about edge-based intelligence and data gathering capability. The deployment of a new meter data management system (MDMS) complemented the AMI deployment to manage data and feed data into other systems across the company.

Fifth, the completion of the deployment of over 100,000 smart thermostats completed a demand response system able to manage over 100 MW of interruptible capacity.

Sixth, the new state-of-the-art billing system, scheduled to go live in April 2011 as this book goes to press, enabled real-time pricing and time-of-use pricing, as well as prepaid service, Web 2.0 services, mobile device access, sophisticated reporting and data analytics, and new services such as solar billing and electric vehicle billing.

Seventh and last, the deployment of a DMS, using many sensors across the distribution grid, integrated to the SCADA/EMS system, integrated with an upgraded GIS, complemented the existing AMI system and completed the first generation smart grid.

Going forward, expansion of the smart grid will involve deployment and management of new systems. The road map included smart appliances, smart electric vehicle charging stations, distributed generation management systems, energy storage management systems, local area networks, and home area networks.

Envisioning and Designing Smart Grid 2.0

Chapter 5 builds on this case study with the story of a groundbreaking community project that envisioned an energy Internet as a new approach to energy production, distribution, and consumption. The Pecan Street Project emerged as an idea while the smart grid at Austin Energy was still being deployed in September 2008, blossoming into a full-blown project with the help of Austin

Energy and other community leaders. Chapter 5 describes the emergence of an advanced smart grid vision from both the lessons learned during the 7-year development and deployment of the nation's first utility-wide smart grid at Austin Energy and throughout the Pecan Street Project in 2009.

Endnotes

[1]   www.theopengroup.org.

[2]   http://www.pmi.org/.

[3]   http://www.metricstream.com/pressNews/pressrelease_63.htm.

[4]   http://www.electrictechnologycenter.com/.

[5]   http://www.lppc.org/index.htm.

[6]   http://www.energy-insights.com/eec/index.jsp.

[7]   http://www.utc.org/.

[8]   http://www.gridwise.org/.

[9]   http://www.nist.gov/index.html.

[10]   http://www.energy.gov/.

[11]   http://www.fcc.gov/

[12]   ERCOT reorganized from alignment around a handful of large zones to a nodal approach with many more small nodes, in order to provide market incentives, align transmission costs and encourage power resources to be located closer to the load they serve. The wholesale market realignment in Texas required all utilities in Texas that were active with ERCOT to adapt.

5

Envisioning and Designing Smart Grid 2.0

In Chapter 4, we thoroughly documented the significant progress achieved by Austin Energy (AE) over the past decade with its smart grid effort, but AE management and staff demonstrated national leadership in a variety of programs, from green power to energy efficiency to smart thermostats to its pioneer smart grid. With its Green Choice program, for instance, AE acted as the principal driver to kick start the West Texas wind farm industry in the early years of the new millennium. More recently, for the *eighth year in a row*, AE was named the number-one utility in sales of green power for 2010, when measured by total kilowatt-per-hour retail sales.

But there are risks in being a pioneer. Austin Energy's solar PV rebate program has historically offered some of the most aggressive rebates nationwide, but its success as a pioneer led to a problem encountered mostly by mature rebate programs: adjustments to the program were required in 2009. As the program grew ever more popular, it became oversubscribed and needed to be recalibrated to accommodate changes in federal tax incentives and the maturity of the local industry. The Power Partner smart thermostat program has so far provided over 100,000 residential and commercial customers (approximately 22% of all AE customers) with free digital thermostats in exchange for their permission to cycle air conditioners on and off during the critical peak periods of the hot summer months, amounting to over 100-MW capacity, making it one of the largest and most successful DR programs nationwide.

With a little help from its friends, Austin Energy started the Plug-In Partners program, which became a national movement that brought cities around the United States together with pledges to add plug-in hybrid electric vehicles (EVs) to their city fleets, providing major car manufacturers the confidence

119

they needed to commit to new EV manufacturing goals. Finally, Austin Energy has completed the replacement of its entire meter stock—410,000 meters—with new smart meters, making it the first utility in the nation to have an operational AMI network throughout its entire service territory integrated to the rest of the grid. As with the solar PV rebate program, there is a price to going ahead of the pack, as AE's smart meters are of an earlier generation and are not capable of communicating energy consumption data directly to in-home devices, as has become standard in the industry.

Introduction

Chapter 4 described the foundation of Austin Energy's original smart grid in a rational IT infrastructure, a smart grid architecture, and an integrated IP network communications network. In this chapter, the trend lines outlined in Chapter 4 lead to the Pecan Street Project, a landmark research program that provided valuable insights on transitioning from an application-led smart grid (1.0) to a network-led smart grid (2.0). Smart Grid 2.0, an integrated, advanced smart grid, will provide innumerable advantages over the conventional approach to smart grids, not the least of which will be lower total cost of ownership (TCO), more rapid deployment, and more flexibility to accommodate unexpected changes in the future.

This next stage in the smart grid journey reassesses the vision of Smart Grid 1.0, where the goal remains the same—the transformation of the energy utility from its roots as a centralized, fossil fuel-driven analog power grid—but the question asked is different. As technology progresses and as energy users gain new awareness, integration of new elements with the smart grid will become ever more essential, suggesting that the smart grid should be designed from the start with robust future needs in mind, such as integration. The Pecan Street Project, which we'll explore in great detail in this chapter, is based on the belief that significant benefits are possible through integration—of DER systems, transportation and water—on a common, more resilient smart grid infrastructure.

Few utilities so far have paused to look at the entire universe of strategic options; the Pecan Street Project may well be unique in its bottom-up, inclusive, community-oriented approach to visioning and strategic planning. From the start, the Pecan Street Project charted new territory, expanding the smart grid focus well beyond DA and AMI, leveraging the significant experience embedded in previous efforts at AE to envision a plan to build a network that could fully integrate dynamic *distributed energy resources* including water and transportation infrastructure, to help the utility achieve sustainability.

This key insight on the importance of IP networking as a new approach to smart grid came as the idea of the Pecan Street Project was first floated in mid-2008. Austin Energy management embraced the opportunity to participate in a broad-based community effort to help it in its journey to reimagine the provisioning of energy services for the coming decades and to explore a network-driven smart grid transformation. The community leaders who launched what came to be called the Pecan Street Project envisioned clean energy as the foundation of new economic development in Austin that would also provide energy security and environmental health over the long term for Austin citizens. Austin Energy managers saw even more. Throughout the Pecan Street Project, Austin Energy managers gave their support in leadership and man-hours to make the Pecan Street Project a success, but also drew valuable lessons from their interactions with industry, academia, nonprofits, and community volunteers. And the effort paid off in helping Austin Energy determine a new future to match its significant progress over the past several years in creating its Smart Grid 1.0. In building its pioneer Smart Grid 1.0, AE positioned itself to learn even more from the Pecan Street Project, which provided a second bite at the apple, an opportunity to refine lessons and craft a new, more effective approach.

The Pecan Street Project: A New Approach to Electricity

In 2008, a group of Austin civic leaders began meeting to discuss an intriguing proposition: Could Austin repeat their early economic development success in computers and semiconductors in the emerging field of clean energy? Two notable Austin high-tech successes in the 1980s became models for economic development success. The Microelectronics & Computer Technology Corporation (MCC), formed in 1983, was a research consortium financed by 12 technology companies to promote research in supercomputers and related technologies. Similarly, the SEMATECH semiconductor research consortium launched in 1987 as part of a strategy to preserve national competitiveness and competency in semiconductor chip manufacturing became a local economic development engine over the next two decades, ensuring Austin's position as a global center for semiconductor manufacturing. And moving beyond semiconductors, SEMATECH would help burnish Austin's high tech reputation, leading to the launch of local tech startups in the Internet boom that followed a decade later.

City leaders pondered whether the creation of a clean energy consortium could similarly attract established clean tech companies to Austin and also help incubate clean energy start-ups. After all, in the emerging clean energy economy, regional economic growth has become inextricably linked to technology innovation, environmental health, and green job creation. Thus, Pecan Street Project came to be about much more than economic development.

Why the name "Pecan Street?" In the early nineteenth century, when city planners laid out a transportation grid for the new capital city of Austin on the north bank of the Colorado River, they named the North–South streets after the rivers in Texas, and the East–West streets after the trees of Texas. Halfway up from the river to the Capitol building ran the main East–West artery, named Pecan Street. In late 2008, when city leaders brainstormed a name for their new community clean energy project, they looked for something distinctly Austin and finally chose to name the project after the original Pecan Street, now widely recognized as Austin's Sixth Street, home to one of its burgeoning entertainment districts. Thus, a nineteenth-century transportation grid lent its name to a project that would provide key insights into a twenty-first-century energy and information grid.

The Pecan Street Project's inclusive, community-led approach reflects in many ways the central role that Austin Energy plays in the life of the Austin community. Although Austin has grown into a leading metropolitan area of the state and nation, it still has a small-town feel to it. Tackling smart grid planning in an inclusive, community fashion was very Austin-like. Volunteers from large IT corporations (Cisco, Intel, GE, Oracle, IBM, Freescale, Dell, Microsoft, Applied Materials) and local small business innovators (Xtreme Power, Helio-Volt, Austin Technology Incubator, Greater Austin Chamber of Commerce), together with world-class researchers (University of Texas, SEMATECH), environmentalists (Environmental Defense Fund), and one of the most progressive energy utilities (Austin Energy) comprised the teams. A key benefit of being so inclusive was to harness innovation and creativity in a friendly, open environment. Leaders hoped the diverse mix of volunteers they assembled would produce fresh thinking and insights not only on timely, critical issues facing the industry, but also on thorny, old problems that have bedeviled the electric utility industry since its inception.

A New Design, Business Model, and Empowered Energy Consumer Class

The utility business model launched by such early electricity pioneers as Edison, Westinghouse, Tesla, and Insull in the late nineteenth century made electricity reliable and affordable and drove economic growth for over 100 years [1]. But now the grid faces significant disruption, challenged by the need to reduce reliance on fossil fuels, but also by the rise of increasingly viable technology-based alternatives to grid power. As an early adopter community, Austin is at the forefront of this wave of change. Its citizens are eager to adopt the latest technologies and naturally gravitate towards the edge of change. However, every energy dollar a utility customer saves by applying new technologies becomes a revenue dollar lost for the citizen-owned energy utility. Technology integration will be on the critical path to infrastructure reform, but so will the question of a new, business model that can more readily adapt to a steady evolution in technology-based energy solutions.

Community leaders in Austin realized that they must approach these critical changes in deliberate fashion, since significant success with alternative energy would risk undermining the utility's financial foundation, even as it *increased* demands on services only the utility can currently provide, including construction and maintenance of transmission and distribution lines, backup power, and reliability. To compound the financial risk inherent when declining revenues meet increasing costs, the city has also come to rely on Austin Energy revenue transfers to fund city services—as a city department, AE transfers roughly $100 million each year from its profits to Austin's general revenue fund.

The original mission of the Pecan Street Project then was threefold. First, it sought to *reinvent the city's energy, water, and transportation systems* through integration of the most advanced technologies and systems, while maintaining financial and environmental sustainability. Second, it would foster the creation of *new clean energy industry companies and jobs* in Austin. Third, it planned to provide *a replicable model for systemic change* for other communities in the United States and around the world.

Other municipally-owned utilities, investor-owned utilities, and cooperatives face a similar transitional challenge: like Austin Energy, they will need a new sustainable business model that will support the transition to a twenty-first -century energy economy, making full use of advanced DER technologies while maintaining the infrastructure necessary to satisfy a growing demand for reliable electricity.

The twin challenges of DER integration and business model transformation combine to make what some social scientists have labeled a *wicked* problem [2], a problem characterized by complexity and multiple subproblems that lacks clear solutions or right answers. Unlike lesser problems, wicked problems are not so much solved as they are whittled down into smaller bits, to make them increasingly more manageable. Collaboration among a broad group of stakeholders enables iterative problem solving by way of pilots, flexible platforms, and experimentation that increases clarity and coherence over time.

Another way of looking at the Pecan Street Project, then, would be as an assessment of the challenges it faced during the transition to a new energy ecosystem: *technology integration* and *business model development* on the one hand—the wicked problem—combined with the third great challenge, *community engagement*—the solution to the wicked problem.

Starting with Strategy

Undertaking a study of an integrated complex system like an electric utility requires a strategy. The Pecan Street Project organizing committee met and determined their strategic approach would be to divide their objectives, tasks, and volunteers into logical groups, and then throughout the first 6 months, adjust and recombine the groups as indicated. From the long view, the planners also recognized that accomplishing the original goals would come in phases.

As Phase One, a community-wide objective review and discussion of strategic issues regarding the creation of an energy Internet, wound down after July 2009 when the teams quit meeting, the project planners acknowledged a need to ratify the *subjective* Phase One conclusions that we'll focus on in this chapter, with *objective* conclusions to be based on the more quantitative work to follow in Phase Two. Phase Two began with a $10.4 million ARRA Demonstration Grant award from the U.S. Department of Energy in November 2009. With a few adjustments, the steering committee from Phase One reconstituted itself as the board of directors of the new nonprofit Pecan Street Project Inc., formed to pursue more quantitative analyses, pilots, and demonstration projects in Phase Two.

Change on Three Dimensions

Phase One told three storylines, aligned with the challenges enumerated earlier. First, the story was about *technological change*, more specifically, the five essential components of DER (energy efficiency, demand response, distributed generation, electric transportation including plug-in vehicles, and energy storage). The integration of water infrastructure and integration of the separate DER components into a cohesive whole called the smart grid completed the technology change storyline.

Second, the story was about the *impacts* of technological change and the *requirements* of such massive integration, leading to a discussion of options for redesigning the rate base and alternative sales of electricity services, collectively referred to as *business model transition*.

Finally, the story would need to consider critical societal elements and the *community engagement* needed to accomplish the requirements on the technology and business fronts.

Getting Organized

Drawing on both the local community and experts from academia and industry, the Pecan Street Project ultimately organized some 200 volunteers, experts in the fields of energy, telecommunications, software, hardware, project management, policy, finance, behavioral change, water, and sustainability (see Figure 5.1). The volunteers were placed into a dozen teams to brainstorm the broadest array of possibilities according to a common strategic outline. In Phase One, the 12 teams met weekly over a 7-month period and ultimately generated hundreds of discrete ideas. It soon became apparent to the steering committee that the value of the project would come from both the detailed assessment of different options in each of the team categories, and from a detailed assessment of DER integration using the connective fabric of the evolving smart grid. As the project progressed, the complexity of the undertaking began to unfold, given the increasing pace of innovation in a variety of clean energy segments, a variety of changes on the policy front, and the critical need to keep the grid functioning during any transition.

While nine of the 12 teams dove deeply into the particular arcane details of their 6-month study plans, three teams were more concerned with integration of the project scope into a cohesive whole. Team 7, the Operations, Systems Integration and Systems Modeling Team, soon labeled "the Smart Grid

Pecan Street Project

Teams

1 Distributed Generation

2 Energy Efficiency

3 Demand Response

4 Transportation

5 Water

6 Energy Storage

7 Operations, Systems Integration and Systems Modeling (Smart Grid)

8 Business Model

9 Customer Interface

10 Regulatory and Legislative

11 Economic Development

12 Workforce Development

Figure 5.1   Project teams.

Team," tackled the issue of technology integration, system redesign, and system modeling. Team 8, the Business Model Team, grappled with the financial and economic implications of the monumental changes facing the utility industry. Team 9, the Customer Engagement Team, focused on previous surveys conducted by Austin Energy and work at the national level to understand and integrate customer energy use and behavior into the new energy paradigm.

During the course of the 7-month process, each of these three teams had individual meetings with the more detail-focused teams. In June, as the process wound to a close and reports were being written, the 12 teams were reconstituted into four "super teams" focused on key DER issues—energy efficiency, demand response, distributed generation, and EV and members from Teams 7, 8, and 9 worked more closely within the super teams to provide further integration. Throughout the Pecan Street Process, Team 7 remained the team with the greatest focus on system design and modeling, on the transformative technology and on the task of integration with a new more resilient version of the smart grid; what we now call Smart Grid 2.0 or the advanced smart grid.

Operations, Systems Integration, and Systems Modeling: Team 7

Team 7 met about 30 times between February and August 2009, establishing a strong task orientation based on effective communication, shared workload, and healthy group interaction. Each week, Team 7 met at Austin Energy headquarters in downtown Austin, employing Web-based meeting tools for those who preferred to join remotely. Team 7 became something of an umbrella organization during the Pecan Street Project, pulling together the work of the other teams in the Pecan Street Project within the Pecan Street Architecture Framework it developed [3]. To create a comprehensive, integrated new electricity paradigm, Team 7 continuously refined the Pecan Street Architecture Framework and related systems, technologies, and integration points.

From 2003 to 2009, Austin Energy had evolved its operations and systems to build the nation's first smart grid in a major U.S. city. The Pecan Street Project offered Austin Energy the opportunity to explore alternatives for expanding on the original smart grid focus of internal IT system redesign and support of individual systems like demand response and distribution automation. The team was able to envision a more complete transition that would prepare the organization and the city for the system integration required by the current rapid advances in network and digital technology, as well as the disruption that was sure to follow. After all, the introduction of new technologies into a distribution grid, while transformative, can also be extremely disruptive, as the following forward-looking case study illustrates [4].

Use Case: Influx of EVs

In March 2017, when Austin hosts the 30th anniversary of the unique South-by-Southwest Music/Film/Interactive Conference, the potential for disruptive impact embodied in the thousands of EV owners who come to town as part of the hundreds of thousands of attendees will become reality. Without prior planning and programs in place, specific distribution feeders would risk being overloaded by the massive influx of large charging loads, energy bills for those whose outlets were used would be dramatic, and charge stations would be difficult or impossible to find in certain areas, while other charge stations would sit vacant and unused.

No electric utility today is prepared for such an influx of mobile, unpredictable load. However, thanks to their Smart Grid 2.0 system and substantial forethought, Austin Energy manages to weather the storm brought on by the massive influx of EV enthusiasts. In advance online, or logging into onboard navigation systems, drivers tap in their destination of "SXSW" and their computers, phones, or EVs are connected to Austin Energy's advanced smart grid and begin the planning cycle with the user. On registration and account creation or reactivation, the smart grid obtains the ID information for the EV, noting such unique descriptors as battery type and capacity and charging capability. Ideal locations for charging and unique pricing arrangements are negotiated based on the user's schedule and profile preferences, as well as the utility's load parameters. Charge rates reflect not only the time of the day, but also grid congestion, and location in proximity to common high load areas. Special rates are assigned for charging stations at the ends of the light rail system, for instance, to spread charging load into the residential districts. Load flow of the grid will drive charge rates at different stations, putting a premium on dense feeders near industrial and large commercial loads.

While EV loads promise to stress the grid in new ways as described in the use case above, the potential for disruption is not limited to EVs—the entire DER category deserves more detailed attention. From the perspective of the Pecan Street Project and for our purposes, a closer look at the term *DER* reveals the following elements [5].

1. *Energy efficiency* (EE) begins with the building infrastructure and the energy appliances included inside. EE may often be low-tech, but can also incorporate technology advances to seal the building envelope and replace old inefficient energy appliances with newer, more efficient models; *residential* EE includes such advances as smart appliances, in-line water heaters and solar thermal technology, redesigned efficient windows, radiant barriers in attics, and spray foam insulation; and *commercial* EE includes building energy management systems (BEMSs) and much more.

2. *Demand response* (DR) on the residential side is only now emerging, but on the commercial side, DR has become somewhat mature. DR is facilitated by Smart end devices/appliances located beyond the meter, which includes both home energy management systems (HEMSs) for residential DR and BEMS for commercial DR.

3. *Distributed generation* (DG), is primarily rooftop solar PV, but also includes microwind, combined heat and power (CHP), geothermal, and any new invention that fits the mold of smaller, more numerous generation plants closely aligned with the load they are intended to serve, rather than providing power to be distributed long-distance over a grid.

4. *Energy storage* is divided in two groups, mobile and fixed. *Electrochemical* storage (e.g., batteries) is available as *mobile* energy storage (e.g., plug-in electric hybrids and pure-play EVs) and *fixed* energy storage, which includes utility-scale storage and community energy storage (CES). *Thermal* storage can be either *cold* thermal energy storage (e.g., ice machines to make ice and supply chilled water to HVACs and chillers, or using HVACs during the day to prechill homes) or *hot* thermal energy storage (e.g., solar thermal devices that store heat in water and/or melted salt).

The Pecan Street Project and Team 7 in particular intended to provide a road map for incorporation of all these aspects of DER into the utility's operations and systems to transform its potential, while mitigating and managing any disruptive effects, both leveraging and evolving the smart grid.

Pecan Street Architecture Framework (PSAF) Design

To understand the design of a future integrated energy delivery system, Team 7 started with a review of the traditional electric utility supply chain and composed a current state architectural framework (CSAF) for the utility environment. The framework is characterized first by *domains*, as follows: (1) *electricity generation* from turbines driven primarily by steam and to a lesser degree, water, (2) *wholesale energy operations*, providing for the provisioning of individual electricity distribution systems, (3) *electricity transmission* over long-distances at high voltage levels, (4) *electricity distribution* in local areas at lower voltage levels, and (5) *retail energy operations* including metering, billing, and customer service.

These are the traditional five domains that make up the energy ecosystem from an operational and systems standpoint. A key challenge for the electric industry in the coming years and decades will be to adapt this fundamental supply chain to future requirements by adding DER as a sixth domain and to enhance the transmission/distribution grids by transforming them into advanced smart grids.

Central to the task of creating a new framework to remap these domains is a paradigm shift in the vision of an electricity distribution system. From the *traditional* perspective, the principal task of the network system has always been to distribute centrally produced power down the line in a one-way flow, out to the edge for consumption, where the electrons are used to accomplish a variety of work tasks by producing light and powering appliances. From the new *transformational* perspective, the principal task of the network system will be to distribute power produced from a much more diverse pool, not only from large remote centralized power plants and medium-sized intermediate power plants, but also from much more numerous, but much smaller distributed power plants located out at the edge and designed to be near the load they serve, which necessarily leads to the risk of power flowing back from the edge into the grid when the distributed plants produce more than can be consumed out on the edge.

Further, this new perspective must also incorporate two other elements rarely incorporated in utility strategic planning today: first, demand response—on-demand curtailment of energy consumption—at a far greater level of integration with all types of customer levels, and second, both fixed and mobile energy storage, which dramatically changes how and when energy can flow and dramatically alters the economic equation so central to utility planning and operations (in essence, the traditional paradigm has depended upon just-in-time energy production to match energy consumption). In short, the new paradigm for energy distribution can be described as *two-way flow of both energy and information* in dramatically different ways than the original power distribution design ever encompassed.

Team 7 began its work by crafting a new framework, dubbed the Pecan Street Architecture Framework, which would include new processes, system approaches, relationships and technologies, thereby extending Austin Energy smart grid efforts to cover the new Pecan Street Project focus. Putting together a model for future integration proved a significant challenge, as it needed to be robust enough to incorporate multiple DER elements as they are developed and deployed while managing transitional business model issues. Essentially, the integration model would need to constitute a dynamic plan for maintaining critical operations and infrastructure during a transition to an increasingly hard-to-predict future. The flexible, iterative model would need to outline a means to progress from an existing layer of technology—the established *Smart Grid 1.0* platform developed and deployed over the last several years—to a new platform that could be called *Smart Grid 2.0.*

Key elements used to drive such planning in Phase One of the Pecan Street Project included not only last mile communication and adoption rates of DER elements, but also development and adoption rates for energy storage, which provides such a dramatic transformation that it should be called out from the other DER elements. Phase Two analysis and demonstration projects would be used to test assumptions developed in Phase One to establish quick wins, identifying potential economies of scale and successful deployment methodologies. Smart Grid 2.0 design needed to emphasize efficient investments, and when sufficient funding existed, recommended investment in longer term programs. Smart Grid 2.0 planning needed to accommodate a range of goals in the short, medium, and long term.

In summary, the role of Team 7 was to extract necessary functionality from group discussions and to use homegrown integration and design tools (e.g., the self-developed system architecture matrix), as well as external tools like the final NIST Interoperability Smart Grid Roadmap to map the elements that make up each domain of the Pecan Street Architecture Framework (PSAF). By comparing current solutions with requirements developed during the analysis phase, the team identified gaps to fill in order to transition the utility to a new state and meet Smart Grid 2.0 needs, most notably providing for effective DER integration.

Using the NIST document, for instance, a review of the different gap categories identified by Team 7 shows the following:

1. *Adequate cyber security.* Current dedicated application networks (AMI network, DA network, DR network, and so forth) lack the capacity for multilevel/multilayer security from end-to-end devices, to network, to the utility network operations center. The elements of a robust security system for Smart Grid 2.0 must include smart device identity, secure digital keys and certificates, secure authentication and encryption, secure communication, and secure data transmission. Such a system would leverage sophisticated hardware and software with multilevel/multilayer solutions, allow for discovery of the device at the software and hardware layer using public and private keys, be capable of authentication and authorization via the network layer and verification by the utility control center, all with secure encryption and replay protection.

2. *Bandwidth connectivity.* Smart Grid 2.0 requires a future-proof communication network—an IP network that can provide connectivity throughout the service territory at speeds and capacities necessary to manage electricity in the new environment. Real-time data transfer capability will be needed for mission critical control systems (from 20 to 100 milliseconds), to support emergency situations, as well as real-time resource management and dispatching. And the exponential increase in data traffic that can be expected based on the proliferation

of devices throughout the utility's service territory must be accommodated (minimum capacity of 2 MB).

3. *Dispatch scheduling.* Today, distribution utilities lack the engagement, connectivity, cyber security, and pricing rules needed to maintain system reliability *and* enable third-party DER owners to be significant participants in the new energy ecosystem. These new parties and new resources need to be able to engage and disengage with the grid on a near automatic basis. The number of transactions inside a DER-enabled distribution grid far exceed current capacities and not only drive the need for more communications bandwidth (i.e., IP networking), but also a new set of rules and standards for distribution grids modeled on those currently practiced in transmission grids nationwide.

4. *Standards and interoperability.* The current distribution grid operates under legacy standards and proprietary equipment, whereas Smart Grid 2.0 will need to be able to incorporate a variety of vendors, applications, and hardware, and ensure that they will all interoperate according to the old standards (DNP3, MUD-BUS), as well as the new standards (IEC 61968, IEC 61850, and IEEE 1547).

In close collaboration with the other Pecan Street Project teams, Team 7 expanded its vision beyond gap identification to include new processes, system approaches, and relationships, ranging from alternative forms of generation to smart grid distribution upgrades, the leveraging of smart meter functionality, and dramatic new potential on the demand side of the grid. At each step of this dynamic process, new technologies, systems, and integration points will facilitate efficiencies and enable cleaner production, distribution, and consumption of energy.

At this stage, it makes sense to pause and walk through the process in more detail, showing how the process unfolded chronologically and how tools emerged for the problems at hand. After all, the processes and tools Team 7 developed and used in Phase One provide a road map and model for other utilities with similar challenges, to help managers explore and define a path for accomplishing the necessary changes to operations and systems and identify options for the utility and the community.

As stated previously, the team started by creating the PSAF with detailed discussion to determine how to define and organize the domains and subdomains for effective representation of the Austin Energy supply chain. First, the team used *mind mapping software* to project a graphic on the wall and over the Web, which allowed them to interactively diagram the domains and subdomains that comprise the local energy ecosystem and produce the PSAF. The new framework not only documented the current utility supply chain, from energy generation to consumption, but also provided a template to guide the integration of distributed energy resources (DER). The following section is an excerpt from the Team 7 notes from the day in February 2009 when the PSAF was laid out.

Power Engineering Concept Brief

The domains in the Pecan Street Architecture Framework are listed below, with subdomains highlighted. One of the first tasks Team 7 took on was to map these domains using mind mapping software, as described above, to produce the Pecan Street Architecture Framework (PSAF).

Domain 1: Central Generation

Team 7 determined that Domain 1 would not be included in their project, focusing their attention instead on more in depth discussion and exploration of the remaining domains. Subdomains in this domain include: Subdomain 1: ERCOT Grid; Subdomain 2: Power Plants; and Subdomain 3: Wholesale Market.

Domain 2: Generation Market Operations

This domain features two subdomains: Subdomain 1 (Retail Market Operations) is defined as "the area under AE control," while Subdomain 2 (Resource and Generation Management) is defined as on/off and any other load control is the point at which other PSP teams will link in their work product (Teams 1 (DG), 3 (DR), 4 (EV), 6 (ES) linked to this subdomain).

Domain 3: System Operations

The four subdomains include: Subdomain 1: Transmission; Subdomain 2: Distribution; Subdomain 3: Wire Field Operations; and Subdomain 4: Control Ops. Key questions posed in this discussion included: "Would the utility need to own new emerging assets inserted into the distribution system, beyond current assets?" and "Who would manage and maintain those assets? Would this be a new service and source of revenue for the utility?"

Domain 4: Metering

This domain features two subdomains: Subdomain 1: Meter & Associated Field Systems; and Subdomain 2: Meter Field Operations. "Communications and Data" should apply to both subdomains. Where the meter fits in the business process would become one of the key questions in Phase One. Emerging legal issues and business model constraints would need to be examined. Customer metering functions are seen as a part of the smart grid, but revenue metering should be distinguished from allowing remote control of appliances. AE went live with a new metering system in August 2009, expecting a new digital billing system to go online by April 2011, making time of use pricing (perhaps even real-time pricing) technically possible.

Domain 5: Distributed Energy Resources

The four subdomains include Subdomain 1: Energy Storage (plug-in hybrid vehicles (PHEVs) are designated as a subset of storage); Subdomain 2: Demand Response; Subdomain 3: Distributed Generation; and Subdomain 4: Energy Efficiency. A key distinction should be highlighted between demand response (DR) and energy efficiency (EE). Energy efficiency is passive (nonintelligent and nonresponsive) while demand response is interactive (it can be monitored and controlled via communications in the smart grid).

Domain 6: Customer

This domain is different from the five that precede it: all the other domains speak to physical systems and design, but this domain talks about users/customers. This domain would become quite crowded after much discussion, with specific definitions added for all who use the electricity produced and distributed by Austin Energy (beyond the traditional residential, commercial and industrial typical in utility systems)—Subdomain 1: Commercial; Subdomain 2: Residential; Subdomain 3: Industrial; Subdomain 4: Builders (green developers); Subdomain 5: Energy Providers (companies that will emerge to offer new retail energy services); and Subdomain 6: Government. Regarding Builders, it was noted that a new set of codes would be needed for builders/developers to incentivize and create new architectural concepts.

PSAF as Integration Tool

The PSAF thus served as the graphic representation of the domains and subdomains that constitute the supply chain of Austin Energy's market elements, infrastructure, and systems. The PSAF also became a tool to integrate the work of the other Pecan Street Project teams and served as a graphic representation of system integration. The PSAF diagrammed the systems and technologies and integration points that comprise the new paradigm, in both the current state and the future state (visionary), from operational, systems, and technology perspectives. Put another way, the architectural framework, with its components of domains, zones, stacks, and integration dimensions, outlines the systems that comprise the architecture and addresses the technologies that enable those systems. While *domains* are the major groupings within the framework and *zones* are the means to divide the utility's operations and functions, for example, the retail operations and wholesale operations. *Stacks* are the elements of interaction, including people (employees and customers), software, hardware, communications, security, and energy. Finally, *integration dimensions* are touch points between the domains.

Next, Team 7 began discussion using predetermined *initial questions* provided by the Strategy Implementation Team and the Governance Board, a process element that all teams followed. Team members were asked to provide individual answers to the questions, which the project manager collated and analyzed, creating the kernel for initial group discussion and providing valuable team-building exercise. After 2 months into the process, Team 7 members had completed the questions and answers, augmenting the list as more questions arose in discussions, finally doubling the original set of questions.

Day-in-the-Life (DITL) Scenarios and Use Cases

The team took another 2 months to develop and discuss *day-in-the-life (DITL) scenarios,* disregarding any previous notions of current technical capabilities and ultimately reorganizing the scenarios by subject area. The value of a DITL scenario lies as much in the development of the scenario as in its contents—it's as much about the journey as the destination. A DITL scenario provides a step-by-step evaluation of a particular change issue, revealing both positive and negative impacts that may have otherwise gone unnoticed until much later.

Next, the team used the best of the DITL scenarios to build nine use cases according to a standard template. Use cases differed from the DITL scenarios as more formal, higher-level evaluations of a potential change in operations or systems. The use cases were compared against the PSAF to ensure that each had the appropriate domains and subdomains in the right places with the right integration dimensions. This collaboration process was used to elicit new elements to the PSAF and help the team to rethink their concepts on current capabilities and what could be possible and when. Throughout the exercise, focus was maintained on desired functionality and convenience for the consumer.

Before getting underway with scenario discussions, however, the team spent some time on exercise design. First, the team saw patterns that would allow grouping the scenarios, for instance, the scenarios described adding devices and/or processes to the grid in three logical groupings. A *resource* primarily describes the supply side, in this case, new DG (e.g., solar PV panel), so these scenarios describe how new generation is treated in evaluating grid options. A *load* on the other hand, describes the demand side (e.g., smart appliance) that is incorporating far greater demand response functionality through such technologies as home energy management systems (HEMSs). Finally, the category *both* described storage (e.g., EV), which is the most disruptive of all, since it can be either a resource or a load depending on where it is in the charge or discharge cycle.

Another way to consider scenarios involves taking either *system* or *user* perspectives. Each scenario would require a description and notation of impacts on other Pecan Street Project teams. These criteria comprised a matrix/template for future discussion. Next, solution scenarios divided along two major alternatives: a *simple tech* solution would be low-tech, based on rules and processes that elicit the human feedback and behavior changes needed to accommodate a norm. In contrast, the *smart tech* approach would be high-tech, highly flexible and intelligent, characterized by tools and technologies that enable individual events. Scenarios were culled to produce use cases.

Other Smart Grid Planning Tools

The completion of use cases that correlated with the Q&A and the DITL scenarios marked the end of the first part of the team's work, whereby they were able to move on from analysis to integration and synthesis, as shown in Figure 5.2 in detail. To determine the issues and impacts of each of these use cases on the energy and information systems within the local energy ecosystem, the team devised a *content collection matrix*, which mapped each use case onto a *system architecture matrix*. Near the end of the process, the team developed a list of *business ideas and recommendations*, which they then mapped onto an *idea assessment matrix*.

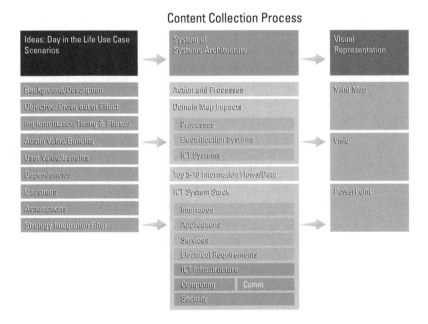

Figure 5.2 Content collection process.

Along the way, the team gathered unaddressed items in a *parking lot list*, which they reviewed near the end of the process, assigning relevant items for further analysis. The team also met periodically with other teams to gather feedback and coordinated with the strategy implementation team along the way. The team submitted monthly *interim reports* to the Governance Board throughout the 7-month Pecan Street Project journey.

Flexible Planning

Team 7 demonstrated a great degree of flexibility along the way to accommodate internal and external changes. For instance, as referenced earlier in this chapter, a great opportunity arose near the end of Team 7 discussions and deliberations when a new resource, the draft *NIST Smart Grid Interoperability Standards Project*, was brought on line as the fruition of earlier industry work. Back in 2007, the National Institute of Standards and Technology (NIST) had been given primary responsibility by the federal government under the Energy Independence and Security Act (EISA) of 2007 to coordinate development of a framework, including protocols and model standards for information management to achieve interoperability of smart grid devices and systems, which was subsequently termed the *NIST Smart Grid Interoperability Standards Project* (also see Chapter 6). NIST will use its report and a supporting organization formed in 2009 to help guide development of smart grid interoperability standards. Team 7 determined that the NIST document would be a good companion to its team report, and after an extensive review, mapped relevant use cases from their report against the NIST draft.

In another instance of rapid adaptation to changing circumstances, the Pecan Street Project planners had to adjust to the Stimulus Bill (ARRA), which was passed only after the Pecan Street Project had launched. Suddenly, with the passage of ARRA, a new opportunity for funding became available, so the team shifted from planning for potential bond issues to planning for potential smart grid grant opportunities. While Austin Energy filed an application for Smart Grid Investment Grant (DOE FOA 58) based on its work with the Pecan Street Project, it was not among grant recipients. But the Pecan Street Project decided in July 2009 to file an application for a Smart Grid Demonstration Grant (DOE FOA 36), and in November 2009, it was awarded a $10.4 million grant that would become the foundation for Phase Two (see also Chapter 6).

The Challenge of DER Integration and Smart Grid Design

Perhaps one of the most significant achievements of Team 7 in Phase One was to provide a much richer understanding of the challenges of integrating DER into the smart grid. The in-depth discussion to answer the questions during the first few months, then the development of scenarios and use cases provided detailed insights on how to make the grid ready to add thousands of EVs and incorporate a vastly larger amount of rooftop solar PV panels than is currently contemplated on today's distribution grids.

The details of DER integration matter a great deal to the utility providing the electricity and ensuring reliability of the distribution grid for all the users, not just the owners of new DER assets. DER challenges must take into account the penetration, predictability, relative distribution and finally, the usage patterns that will determine the level of the integration challenge. The dimensions of challenges presented by DER integration include: (1) infrastructure cost, (2) grid impacts and reliability, (3) safety, (4) communication, (5) codes and regulations, including siting requirements, (6) variability, and (7) security.

The *economic* challenge for utilities will be to devise ways to avoid the inherent threats of DER integration, such as unplanned and irregular daytime charging, clustering, and meager revenues as EVs and other DER become ever more common. Avoiding a blown transformer by adding distribution system capacity or upgrading a substation to accommodate such clustering will likely cost more than the utility could ever earn from the electricity alone that it will sell (consider that utilities will earn relatively little from EVs that charge mostly at night, when electricity is priced by new time-of-use (TOU) rates that may be as low as 2 to 3 cents/kWh), special EV rates may be required.

Current grid design matters immensely. In general, a single transformer in the United States may service only 4 to 10 homes, but in Canada and Europe, which have a different grid design, a larger transformer will serve as many as 20 to 30 homes, often even more.

Transformers and feeders will need to be upgraded and feeder management capabilities will need to be automated to accommodate EVs at any scale in order to minimize *grid impact.* Automating a distribution feeder to accommodate EVs will typically require adding three switches (at an approximate cost today of over $5,000 each). Beyond the distribution substation, protection gear such as relay protectors are not currently present to protect against adverse impacts from EVs.

The utility must work with the emerging EV manufacturers to enable scheduling capability for EVs, so the utility can control these events in real time—which drives the need for IP networking infrastructure throughout the grid. Put another way, just as the early grid was designed to meet peak load on a macro basis, the emerging smart grid must be designed to meet peak load at the edges, not only for safety, but for reliability and economic life of the grid.

As part of the new grid design, for instance, DER integration strategy now must encompass the concept of *islanding*, which would enable the grid to disassociate below the feeder level under certain conditions. Fortunately, DG technologies provide for sustainable islanding. A single control center today does not accommodate multiple islanding scenarios as a regular daily occurrence, so enhanced smart grid control will become ever more necessary.

Harmonics measuring and monitoring and modulation will become much more important to grid managers when integrating DER, not only for grid balancing, but also to ensure long equipment life. *Communication* capabilities of the grid will need to be significantly enhanced. Adding a variety of DER devices with different load characteristics will create a richer, complex management challenge as power quality begins to fluctuate more widely. If the current range of voltage fluctuates between 116 and 124V for an average of 120V, the addition of new devices can be expected to expand that range by a matter of degrees, to say 108V to 135V. Inductive load in particular whether from EV charging stations or solar PV inverters, generates reactive power at far greater levels. And further, solar PV carries the potential to provide an additional burden for the grid, given the *variability* and unpredictability of its electricity production.

Until the grid has been redesigned and enhanced to accommodate significant amounts of DER, however, changes in *codes and regulations, including siting requirements,* are likely to be implemented as a practical way to ensure grid stability and harmony, by prescribing where, when, and how DER elements can interconnect with the grid.

And finally, the *security* aspects of DER must be recognized from the outset. As described earlier in this chapter, security is a key feature of the emerging smart grid, and integrating thousands of new DER elements to the grid carries a significant security risk. Security must be considered at each stage of DER integration.

In summary, DER integration will emerge as a far more complex issue as technology advances provide ever cheaper and more functional solutions to add to the grid. Team 7 discussions during Phase One revealed that the smart grid will be the solution to DER integration, but that the rules of integration are only now being mapped out.

Phase Two: Demonstrating an Energy Internet

Phase One could be said to have officially concluded in March 2010, when the Pecan Street Project issued a set of 39 recommendations in a report now available on the Web site [6]. The purpose of Phase Two shifted from *brainstorming* to generating new ideas and recommendations to *applied research* to quantify grid impacts and evaluate solutions in the field.

A key issue for the nonprofit formed at the outset of Phase Two would be to define how it would add unique value without duplication of effort with other Austin organizations. Potential roles for the nonprofit included: (1) outside analytical assistance, including measurement and verification, (2) public relations for a sustainable energy future, (3) project management for projects that receive ARRA stimulus funding, (4) a platform for fundraising on behalf of universities and university researchers in areas where funding gaps exist, and (5) reaching stakeholder consensus on priorities. Identified stakeholders in the Pecan Street Project process going forward began with the new nonprofit organization itself and also included Austin Energy, Austin Technology Incubator, City of Austin, the Environmental Defense Fund, the University of Texas and other organizations in the social services sector, and private sector organizations and businesses.

Many of the early questions about a role for Phase Two were answered in November 2009, when the DOE announced the winners of federal matching grants to demonstrate smart grid technologies under Funding Opportunity Announcement (FOA) 36. Pecan Street Project was awarded $10.4 million, which was matched by over $14 million, mostly in infrastructure, provided by Austin Energy.

The grant application came together in a matter of weeks in July, with a plan to demonstrate an energy Internet at a New Urban-style neighborhood that had already been partly constructed in a recovered section of downtown Austin, on the site where the former Mueller Municipal Airport had once stood. The new neighborhood—and the Pecan Street Project Energy Internet plan—took the same name. The Mueller neighborhood would become home then to the Mueller Energy Internet project—a microgrid in the center of Austin that would serve as a living laboratory to test new energy concepts.

As the neighborhood project got underway, homeowners began to move into homes that had either been built according to LEED designs or had followed Austin Energy's Green Building guidelines. Either way, living efficiently with the help of clean technology has been a driving force from the beginning for this neighborhood. The residents could be described as pioneers in a new way of living, and are likely to be more open to new approaches than the general population. New urban infill neighborhoods like Mueller offer great potential for smart grid demonstrations.

As planned, the Pecan Street Project Smart Grid Demonstration at Mueller will provide an opportunity to put many of the ideas and concepts developed during Phase One to the test. A demonstration project goes beyond brainstorm sessions, modeling and opinion surveys to assess potential solutions to both complex technical issues and novel social arrangements. The project will include, for instance, a Demonstration House, which will embed in-the-home technologies, tools and applications to provide an opportunity for interaction with the general public. Tracking progress at Austin Energy and the organizational framework of Phase One, features of the project include:

- *Two-way electricity meters* that provide customers with real-time information, including mobile phone access, collect and transmit a range of data, including data from smart appliances located inside the home or office, communicate with in-home/in-store displays and/or local energy networks, and measure energy flowing from buildings onto the grid (e.g., from solar PV panels) and communicate the data both to the grid operator and to local energy networks.

- *Local energy networks* (home area networks or office area networks) with supporting software that provide customers control of energy usage down to the appliance level—even remotely over wireless devices, respond to demand response protocols from the grid operator, manage household smart appliances' energy usage based on electric budget, environmental preferences, variable pricing information and other metrics set by the customer, manage household water usage based on electricity budget, environmental preferences, variable pricing information and other metrics set by customer, integrate variable pricing information, and interact with variable inverters.

- An *environmental dashboard* for larger communities (e.g., neighborhoods, all residences and the microgrid) that integrates the information collected through the local energy network, informing customers about the environmental impact of their energy and water usage.

- *Smart appliances* that integrate with the local energy network, possibly including HVAC and other major appliances.

- *Variable inverters* for homes and offices with solar PV that can be adjusted by the grid operator.

- *Plug-in vehicle charging* and energy management systems that integrate into local energy networks.

- *Utility-level functionality* that manages demand response through a microgrid energy Internet, accommodates, accounts for, and manages two-way energy flow, provides information and instructions to local energy networks, and integrates plug-in vehicle charging with solar and energy storage in the Mueller Town Center garage.

- *Open source design* that promotes replication—the intellectual property technologies and systems developed by Pecan Street Project will be open source and freely available, creating protocols for innovators to test their technologies on the energy Internet and to introduce new products and services onto the energy Internet that meet system requirements and achieve system values (e.g., carbon-free clean energy, reduced water usage).

- *Integration of water/reclaimed water systems* with the deployed energy Internet.

- *Distributed generation* including solar PV (panel or thin film, or both), and possibly solar water heaters; and *storage,* including thermal storage, battery technologies (e.g., lithium ion, lithium iron magnesium phosphate, metal air, and lead acid), and possibly ultracapacitor and fuel-cell systems.

- *Business model testing* to measure functionality with customers, private sector job creation, and utility finances.

Like many of the DOE smart grid demonstration grants, the Pecan Street Project at Mueller will focus on smart grid technologies and DER integration. However, the project is also unique in many ways: (1) a commitment to use open source standards, (2) integration of a water system with a smart grid system, (3) the inclusion of green building technology and the integration of changes to building codes, (4) the commitment to utility financial viability, (5) the inclusion of a high level of affordable housing (25%), and (6) the integration of native landscapes. As this project gets underway, it will provide an opportunity to assess many of the insights gained and lessons learned from Phase One.

Pecan Street Project Lessons Learned

When the Pecan Street Project launched in late 2008, the national economy was still considered sound, the federal stimulus bill was not even imagined, and the field of smart grid projects remained quite narrow. The smart grid industry was poised to burst onto the stage, however, as became apparent with the release of two U.S. DOE ARRA grant programs in April 2009. By October 2009, the Pecan Street Project Phase One technical report was in the first draft stage, and the DOE Smart Grid Investment Grant (FOA 58) awards had been made, earmarking $3.4 billion for 100 Smart Grid Investment Grant projects. One month later, the Smart Grid Demonstration Grant (DOE FOA 36) awards were announced, identifying regional Smart Grid Demonstrations (including the Pecan Street Project) that would share a portion of $600 million in funding. Finally, EPRI, the research organization associated with the electric utility industry worldwide, began announcing a program in 2009 for as many as eight smart grid demonstration projects.

Where do the Pecan Street Project and Austin Energy now fit in the context of the evolving smart grid discussion? In short, these two organizations have blazed a trail, but execution and fortitude will determine if they stay there given recent events. Having achieved more than most electric utilities over the past decade in reinventing itself through incremental changes, progressive Austin Energy is now undergoing a dramatic paradigm shift. With the prospect of moving away from its traditional role of distributing commodity kWh one-way out past analog revenue meters to dumb appliances and relatively passive consumers, Austin Energy has before it an opportunity to start providing dynamic energy services over a two-way smart grid that includes smart meters, smart appliances, and more active consumers. Such a shift will require them to redouble their efforts with new approaches, new attitudes, indeed, even a new language, and in Austin, a rate case is pending for 2012. Whether these changes continue remains to be seen. The paradigm shift, if it occurs, may include changes such as those found in the following. Team 7 brainstormed the following list of recommendations in the final days of Phase One.

Team 7 Recommendations

1. *Distribution system operator (DSO).* Over the next 4 years, as AE introduces a variety of distributed resources onto its grid, its responsibilities and operational functionality will move closer to that of ERCOT, its host independent system operator (ISO). AE may choose to become a distribution system operator, a term used in Team 7 discussions. AE will need to have systems and processes to manage market functions and the flow of energy across the grid, like those at ERCOT.

2. *Independent distributed generation (DG) dispatch.* With the ERCOT region in the process of transitioning to a nodal market, which will account for energy transfers based on specific dispatch from node to node, rules will be needed within the AE distribution grid to define parameters for third-party DG dispatch on the AE distribution network, an activity potentially brought under the purview of the new nodal market. This *independent dispatch scenario* will need to be defined with much greater detail.

3. *DR and DG zonal development.* Demand response (DR) and distributed generation (DG) decisions help determine the strategic vision of the utility. Subdividing the service territory into *DR zones* and *DG zones* would help to ensure optimal distribution so that these resources support the utility's vision and operational requirements. The criteria for zonal siting would need to consider distribution congestion, economic development, and disaster recovery. Managing energy costs in schools, a key public policy issue, would argue to put those facilities at the front of the line to receive subsidies for DG and DR, and the util-

ity would also benefit from owning DG on these facilities to facilitate disaster recovery.

4. *Rates versus information.* Rates based on time of use (TOU), real-time pricing (RTP), and critical peak pricing (CPP) help a utility differentiate its commodity kilowatt-hours by price in order to motivate consumers to shift their consumption to off-peak hours. But if a new paradigm of providing energy services were to be adopted, alternative rates may be more appropriate. *Real-time information,* for instance, would educate consumers about their consumption and its impacts in order to change their consumption behavior, perhaps avoiding altogether the need for rate-based price signals.

5. *Cannibalization and transition.* The realignment of a city-owned utility to embrace DG must be managed to ensure that such an alternative does not cannibalize the utility's revenues and city services that provide key revenue for the city budget. A utility choice to shift capital investment from central generation and distribution facilities to utility-owned DER could delay or avoid altogether private-sector DER investment and utility revenue dilution.

6. *Decoupling.* Decoupling breaks the connection between energy sold and income earned by providing a return on existing capital investment and disincentives for ancillary expenses that may become obsolete in a DER environment. A new rate structure based on decoupling would feature a required component embedded in every customer's bill that covered fixed capital investment and an optional array of charges that associate specific costs with specific services.

7. *Incentives versus mandates.* Mandates carry with them an element of coercion that is out of alignment with a more inclusive approach to energy provisioning, suggesting a shift to incentives. The utility that shifts to providing energy services should ask for new customer behaviors to provide a more efficient and cost-effective energy ecosystem.

8. *Customer segmentation and differentiation.* A shift to TOU rates will impact different customer classes differently. For instance, SMB customers have few to no options to shift consumption from peak business periods that coincide with peak demand periods. Multiple programs will be needed to accommodate different classes of customers and feedback loops will be needed to track the performance of such programs.

9. *Last-mile communication.* Current last-mile communication options are inadequate to provide a complete solution to distribution utilities to communicate sufficiently throughout the service territory. The data

requirements to implement the solutions envisioned in this report overwhelm the existing communication options, suggesting a shift from narrow band and stronger emphasis on energy-dedicated last-mile IP networks.

10. *Standards and interoperability.* The adoption of standards will be required for the full vision of an energy ecosystem to be fulfilled. Interoperability is a requirement for a fully evolved, integrated, communicative advanced smart grid. The utility system will need to integrate all the islands that technology providers create with their proprietary technologies. NIST and other groups will need to continue to drive this overarching goal of full interoperability. The utility will need to become a system integrator, providing the API in the cloud to let distributed solutions work together in a functional ecosystem.

11. *Change management and the economy.* Organizational and infrastructure change is facilitated, even inextricably linked by the presence of economic growth, which provides a utility an opportunity for incremental implementation. Systemic transition can focus on gradual replacement of old equipment and processes with new ones to accommodate economic growth according to a new set of standards.

12. *Energy storage as an asset.* Each type of energy storage—central, substation, community and premise—has potential to open new opportunities for distribution operations. But the regulatory treatment of storage must be resolved for this emerging resource to be effectively implemented and deployed in a distribution utility service territory. Resolution of the treatment of this emerging energy resource in the capital markets is needed for it to become a reliable resource for utilities. Energy storage is limited to pilot scenarios while utilities wait for costs to come down benefits to be proven, and capital risk scenarios to be resolved.

Conclusions on the Next Generation Utility

Moving from its traditional role (selling commodity kilowatt-hours over a one-way distribution grid out through meters and appliances to relatively passive consumers) to a new role (providing energy services over a two-way advanced smart grid and smart devices to relatively more active consumers) constitutes dramatic change and demands new approaches, new attitudes, and new terminology.

Austin Energy has long held a vision to become a next generation utility. The Pecan Street Project went a long way to defining in greater detail what a next generation utility will look like. On the other side of a Smart Grid 2.0 transformation will be a utility not only far more reliant on an integrated fleet of diverse DER systems and a base of efficient, smart consumers, but also no longer subject to the risks of rising fossil fuel costs and rising costs from new carbon externalities. And *integration* will define the next generation utility, integration with its own system (Smart Grid 2.0), with emerging DA, DR, and DER technologies, with its own consumer community, with the transportation infrastructure; and with the water infrastructure.

In hindsight, the documentation in Pecan Street Project Phase One of potential reform ideas and new insights and processes to help facilitate change helps to move our national discussion forward. Phase Two promises to reveal even more valuable insights as do the other ARRA Demonstration Grant projects. The devil will be in the details, as they say, and each electric utility will have unique issues to resolve. Austin Energy served itself well with the Pecan Street Project.

In a poll released in September 2009 and again in 2010 [7], *Intelligent Utility* magazine and *IDC Energy Insights* ranked Austin Energy at "near genius" level, as the second-smartest utility in the nation, behind only Sempra Energy of San Diego. Austin Energy has the motivation, the means, and the methodology to remain a leading progressive utility. In the Pecan Street Project, Austin Energy now has a pilot project to evaluate and implement its Smart Grid 2.0 objectives. The challenges AE faces in implementing the many recommendations of the Pecan Street Project discussed in this chapter will not be trivial, but by taking initiative, AE has placed itself in a good position to carry forward with its objectives.

Chapter 6 will provide a review of the current state of smart grid activity in the United States at the time of this writing. We'll explore the advances in research and development, and show how in some cases projects are progressing along the road just described in this chapter.

Endnotes

[1] Thomas Edison invented the incandescent lightbulb or, more succinctly, discovered a filament that would last for a very long time. [Edison also preferred a direct current (DC) system of power plants located close to the load, so we may need to start recognizing him as the "father of DG" at some point.] George Westinghouse favored alternating current (AC) and recruited Nikola Tesla away from Edison's Menlo Park laboratory to coinvent AC generation, transmission, and distribution system design with step-up and step-down transformers that we still have today. Tesla also invented the electric motor some years later to expand the purview of early utilities from electric light companies to the power and light title we are so familiar with now. Samuel Insull, Thomas Edison's assistant, went

on to raise tremendous amounts of money and create a business model to finance the construction of power plants and electric grids: holding companies or "trusts" owning regional investor-owned utilities. His system of interlocking trusts drove an economic boom in the 1910s and 1920s, but then Insull became Public Enemy Number One when FDR was elected. FDR's Attorney General made Insull the first example of his "trust busting," leading to passage of the 1935 Public Utility Holding Company Act (PUHCA).

[2]   http://cognexus.org/id42.htm.

[3]   The Pecan Street Architecture Framework reflects the lessons learned in the Smart Grid Architecture design process described in Chapter 4.

[4]   This EV use case is the inspiration for the similar use cases outlined first in Chapter 3 and later expanded from a different perspective in Chapter 7. The disruptive potential of EVs led us to look at this subject from different perspectives.

[5]   As interpreted in the Pecan Street Project, the term distributed energy resources (DER) included both the relatively low-technology energy efficiency resource class and the higher-technology of demand response. EE and DR were included because of their potential to lower total energy demand requirements and peak demand requirements with a focus on the built infrastructure, which consumes 70% of the nation's electricity, and energy consumption behavior. Buildings built before 1970 are notoriously inefficient, lacking in basic insulation and other energy efficiency fundamentals. Elsewhere in this book, DER includes DG, EV, and ES, but not EE or DR.

[6]   http://www.pecanstreetproject.org.

[7]   Results from the 2010 poll were published in the January/February edition of *Intelligent Utility* magazine http://www.intelligentutility.com/magazine/article/203209/austin-energy

6

Today's Smart Grid

In Chapter 5, we described how the work at Austin Energy had expanded into a community-wide discussion on the prospects for an energy Internet, leading to the creation of the Pecan Street Project nonprofit organization and the launching of a research and development project in the Mueller neighborhood in Austin. The extensive work at the Pecan Street Project in 2009, and at other pioneer efforts throughout the industry, presaged a broader discussion at the national and international level that emerged in 2010 as what might be called a mainstream meme, where smart grid was commonly written about in mainstream publications and television commercials referenced the term.

Smart Grid Emerges as a Mainstream Meme

This chapter tracks the development of the smart grid concept beyond a small coterie of individuals and industry insiders and the emergence of the term "smart grid" into the mainstream in 2010, providing a snapshot of smart grid at the time of this writing [1]. Smart grid embodies a reinvention of the fundamental infrastructure of the modern global economy, painting a picture on a broad tableau across the globe. So it should come as no surprise that momentum and interest in the topic has gradually increased over the years with the steady progress of technology and policy development, to the point that today smart grid is widely considered a realistic, on-the-ground prospect, even "shovel-ready" in some places. Depending on the speaker and the situation at hand, the term "smart grid" may remain a confusing term with a multitude of meanings. As this chapter documents, the smart grid has taken root, but the future of smart grid is not yet defined.

Smart Grid: The Early Days

The term "smart grid," which was defined by Andres Carvallo on March 5, 2004 (see Chapter 1), was heard mostly in the relatively small circles of the utility cognoscenti, among experts at NREL and other DOE labs, at EPRI, GE, Cisco, and IBM, from industry pundits like Clean Edge, among early members of the GridWise Alliance, and perhaps among such early North American utility innovators as American Electric Power, Austin Energy, CenterPoint Energy, ConEd, Duke, Oncor, PG&E, Salt River Project, Southern California Edison, SMUD, and San Diego Gas and Electric [2]. Of course, early smart grid activity was by no means limited to the United States, although that is the primary focus of this book: notable early movers in smart grid circles internationally included DONG (Denmark), ENEL (Italy), North Delhi Power and Light (India), and Country Energy, EnergyAustralia, and SPAN (Australia). The list of early adopters could go on and on; this attempt to name the early movers and shakers in smart grid brings to mind the old adage: "Success has many fathers, failure, only one."

Until recently, smart grid remained a visionary concept, with potential based on probable advances in technology. Earlier this decade, we may have had confidence that the smart grid era would come, but nobody could describe in detail how the story would unfold, and that remains true today. With a few exceptions, notably journeys of discovery such as the on-the-job training inside Austin Energy documented in Chapter 4, smart grid activity has mostly been limited to research, pilots, and planning until very recently.

Clean Edge discussed the potential of grid optimization in its *Clean Energy Trends 2003* [3], describing how EPRI, U.S. DOE, Cisco, and a handful of utilities including Exelon, ConEd, and Salt River Project had formed the Consortium for Electric Infrastructure to Support a Digital Society (CEIDS) [4]. According to its mission statement, CEIDS was an attempt [4]: "to develop the science and technology that will fundamentally transform the infrastructure to cost effectively provide secure, high-quality, reliable electricity products and services." In short, CEIDS was focused on the smart grid before that term had even been adopted to name the nascent trend. To give credit where credit is due, the vision of EPRI and the founding organizations behind CEIDS has stood the test of time. Their 2003 vision [4]—"A new electric delivery infrastructure that integrates advances in communications, computing, and electronics to meet the energy needs of the digital society"—illustrates how the industry was entirely focused on only modernizing the utility infrastructure, while ignoring the transformation that would occur beyond the meter and into buildings, homes, and electric vehicles.

Among other noteworthy early milestones, we must mention the creation of the GridWise Alliance in 2003 and IBM's early work in the development of the smart grid maturity model, which since 2009 has resided at Carnegie Mellon's Software Engineering Institute.

Founded in 2003, the pioneering GridWise Alliance [5] is a group of utilities, large and emerging technology companies, academia, and representatives from the financial community who came together based on a shared commitment to making the smart grid a reality. The GridWise Alliance has been true to its founding vision, allowing a diversity of perspectives to cooperate to shape the policy discussion and keep legislative and regulatory leaders informed.

Beginning in 2007, IBM focused its early efforts in smart grid by creating the Global Intelligent Utility Network [6], whose founding members included CenterPoint Energy (United States), Country Energy (Australia), CPFL (Brazil), DONG Energy (Denmark), Liander (the Netherlands), North Delhi Power Limited (India), Pepco Holdings (United States), Progress Energy (United States), San Diego Gas & Electric (United States), and Southern California Gas (United States). Recruiting the nonprofit research institute APQC [7] to its cause, the group set to work on its first project, borrowing from a successful best practice in the software world to create a tool for utility managers, the smart grid maturity model (SGMM). With the SGMM (see Figure 6.1), smart grid

Smart Grid Maturity Model[1]

Levels	Descriptions	Results
Level 5: Innovating – Next wave of improvements	New business, operational, environmental and societal opportunities present themselves, and the capability exists to take advantage of them.	Perpetual Innovation Self-healing operations Autonomic Business INNOVATORS
Level 4: Optimizing – Enterprise-Wide	Smart Grid functionality and benefits realized. Management and operational systems rely on and take full advantage of observability and integrated control across and between enterprise functions.	Transformation Real-time corrections Broad reuse VICTORS
Level 3: Integrating – Cross Functional	Smart Grid spreads. Operational linkages established between two or more functional areas. Management ensures decisions span functional interests, resulting in cross functional benefits.	Systemization Repeatable practices Shared information CROSS LOB CHAMPIONS
Level 2: Functional Investing	Making decisions, at least at functional level. Business cases in place, investments being made. One or more functional deployments under way with value being realized. Strategy in place.	Strategy Proof of Concepts MISSIONARIES
Level 1: Exploring and Initiating	Contemplating Smart Grid transformation. May have vision, but no strategy yet. Exploring options. Evaluating business cases, technologies. Might have elements already deployed.	Vision Experiments PROPHETS AND HEROES

[1] Smart Grid Maturity Model (SGMM) Overview, Software Engineering Institute, Carnegie Mellon University (Pittsburgh, PA, 2009)

Figure 6.1 Smart grid maturity model.

teams found a guide that provided valuable help for their smart grid projects, in project planning, quantitative measurement of their progress, and prioritization of options. In 2009, IBM transferred stewardship of the SGMM to the Software Engineering Institute at Carnegie Mellon University [8]. The SGMM has been influential in these early days of smart grid: over 60 leading utilities have leveraged the best practices embodied in the model so far.

Launching Smart Grid

If there is a race to build the smart grid in the United States [9], then the years before 2009 would show Austin Energy finishing the first fully deployed smart grid in the United States and closely followed by Oncor, Centerpoint, and American Electric Power in Texas. If the smart grid is a national race, then the starting gun went off with the enactment of the American Restructuring and Reinvestment Act (ARRA) in February 2009 [10], with nearly $4 billion in federal funds allocated to support matching grants in two categories—investments and demonstrations. With that announcement, utilities, vendors, and consultants began assembling teams and developing their grant applications. As grant awards were announced in late 2009, the emerging smart grid landscape unfolded before our eyes with winners receiving a kick start to their projects and those not winning awards left to ponder next steps, as the promise of reduced risk from federal funds evaporated in a single day. Large and small companies throughout the electric industry have now begun realigning and preparing for dramatic change. As the dust settled at the end of 2009, one thing was certain: the smart grid revolution had officially begun.

Smart Grid Speed Bumps

In 2010, with an abundance of smart grid activity underway, some of the bloom has definitely fallen off the smart grid rose, as the realities, costs, and complexities of building a smart grid have become apparent to increasingly more stakeholders. Consumer groups began to object to the costs of smart meter deployments, challenging the value proposition, predictably starting in California with objections in Fresno and Bakersfield to a massive smart meter rollout by PG&E. Regulators investigated, and a few months later, a similar objection arose among Texas energy consumers to an Oncor smart meter deployment. Regulators began pushing back on utility plans for full recovery of smart grid costs, most notably with a landmark case in Maryland, where Baltimore Gas & Electric stumbled in its efforts to gain approval of a rate case as part of its

$200 million DOE matching grant acceptance process. By the end of the year, another type of challenge had become apparent. If AMI were the automatic first step to building a smart grid at the beginning of 2010, grid optimization and DA applications had become a viable alternative by the end of the year. Shifting focus from smart meters to the distribution system had the double advantage of avoiding direct consumer backlash like that in California and Texas and the potential to more readily make the business case to a more skeptical audience of regulators and consumers.

By the end of 2010, the concept of a smart grid had become far more widespread; smart grid steadily crept into our national discourse as a consumer concept, showing up in mainstream publications like *Scientific American* [11], *Newsweek* [12], and the *New York Times* [13]. Although the smart grid is the most significant change to hit the fundamental electric grid infrastructure since its creation over 100 years ago, the actual implementation of smart grid projects has so far been more limited than the advanced vision we've talked about. Nevertheless, as this chapter documents, the nation has been busy laying the foundation for smart grid projects far and wide.

A key element in the gradual development of smart grid in the United States has been a national discussion on standards, interoperability, and security. The National Institute for Standards in Technology (NIST) [14] stands out as the leader in this area, with its Smart Grid Interoperability Panel (SGIP) [15] and its security documentation [16], developed in coordination with the GridWise Architecture Council [17]. In terms of interoperability, several more standards bodies also deserve our attention, given the fact that a plethora of technologies will ultimately support the drive to update the grid with digital devices and applications.

Smart Grid Perspectives

This rest of this chapter is divided into three major sections that examine different perspectives on smart grid. First, we look at leadership at the national and state level through legislative and regulatory institutions that aim to address smart grid issues. Second, we look at U.S. standards that are likely foundational elements to help manage the complexities of the smart grid. Finally, we map the variety of alliances and industry groups associated with the development of smart grid, including helpful media resources to keep track of smart grid progress. These sections are by no means an exhaustive or complete list of smart grid activities and organizations in the United States, but we provide this review to showcase the diversity of activity and the breadth of interest in smart grid.

Government as a Smart Grid Stakeholder

Federal Executive Influence

Office of Science and Technology Policy (OSTP), Executive Office of the President

In November 2010, OSTP [18] issued a report titled *Accelerating the Pace of Change in Energy Technologies through an Integrated Federal Energy Policy* [19], which provides a road map for the federal government to help transform the U.S. energy system within the next two decades.

Federal and State Legislative Influence

Title XIII, EISA 2007

A discussion on federal legislative influence starts with a review of the Energy Independence and Security Act of 2007 (EISA) [20], specifically, the smart grid section, Title XIII. Of the nine sections in Title XIII, six concern the activities of the U.S. Federal Smart Grid Task Force (FSGTF), established in Section 1303, to include experts from seven different federal agencies [the DOE, represented by the Office of Electricity Delivery and Energy Reliability (OE), the Task Force lead, and representatives from the Office of Energy Efficiency and Renewable Energy (EERE) and the National Energy Technology Laboratory (NETL); the Federal Energy Regulatory Commission (FERC), the Department of Commerce (DOC), the Environmental Protection Agency (EPA), the Department of Homeland Security (DHS), the Department of Agriculture (USDA), and the Department of Defense (DOD)]. Title XIII directs the task force to: (1) produce regular reports on the status of smart grid deployments nationwide (Section 1302); (2) carry out a program to research, develop, and demonstrate smart grid technologies and establish a smart grid regional demonstration initiative focused on advanced technologies (Section 1304); (3) establish a federal matching funds program (Section 1306); (4) submit to Congress a study assessing the effect of private wire laws on the development of combined heat and power (CHP) facilities (Section 1308); and (5) submit to Congress a study on the security aspects of smart grid systems (Section 1309). The three other sections: (1) outline federal policy towards smart grid (Section 1301); (2) direct NIST to set up a smart grid interoperability framework (Section 1305); and (3) provide direction to state regulators on appropriate oversight of utility investments in smart grid (Section 1307). Title XIII got the ball rolling with research, funding, and regulatory direction, but it was the Stimulus Bill (ARRA) 2 years later that really launched major smart grid activities.

DOE National Laboratories

At least six of the DOE's 21 national laboratories have a special focus on smart grid (Figure 6.2) [21]. The National Energy Technology Laboratory (NETL)

DOE National Labs and Smart Grid

- National Energy Technology Laboratory (NETL)
- National Renewable Energy Laboratory (NREL)
- Los Alamos National Laboratory (LANL)
- Sandia National Laboratories (Sandia)
- Lawrence Berkley National Laboratory (LBNL)
- Pacific Northwest National Laboratory (PNNL)

Figure 6.2   Smart grid DOE national labs.

[22] in Morgantown, West Virginia, is working on upgrading the national transmission and distribution system by developing a nationally coordinated grid modernization framework. The National Renewable Energy Laboratory (NREL) [23] in Boulder, Colorado, is the primary R&D lab for renewable energy and energy efficiency. Los Alamos National Laboratory (LANL) [24] in Los Alamos, New Mexico, focuses its smart grid activity on the ITC aspects of grid design, grid control, and grid stability. Sandia National Laboratory [25] in Sandia, New Mexico, has projects in renewable energy storage and solar energy technology [e.g., Solar Energy Grid Integration System (SEGIS)]. Lawrence Berkley National Laboratory (LBNL) [26] next to the UC Berkley campus outside San Francisco, among other accomplishments, is researching EV technology and has developed an open source communication specification that supports automated demand response (OpenADR). Finally, Pacific Northwest National Laboratory (PNNL) [27] in Richland, Washington, brings a special focus to the environmental side of smart grid.

Smart Grid Information Clearinghouse (SGIC) and SmartGrid.gov

DOE set up a process for gathering and disseminating smart grid data. Soon after the ARRA announcements on smart grid funding in early 2009, DOE issued an RFP for a $1 million grant to develop and launch an online Smart Grid Information Clearinghouse (SGIC) [28], which will be complemented by the FSGTF's central database for information on the smart grid and government-sponsored smart grid projects [29]. Virginia Tech won the contract and developed the Web site over the course of a year. DOE requires its smart grid grant recipients (FOA 58 investment grants and FOA 36 demonstration grants described later) to provide information from their projects to the National Renewable Energy Lab (NREL), which will then forward the information on to the SGIC, but the SGIC will also be open to information contributed from other smart grid projects and sources.

DOE-Funded Smart Grid Projects

With the enactment of ARRA in February 2009, the DOE issued two funding opportunity announcements (FOAs): FOA 58 for large-scale investment grants and FOA 36 for smaller demonstration grants.

In October 2009, 100 FOA 58 awards were announced for matching investment grants totaling $3.4 billion in multiple categories, with the lion's share of awards going to AMI projects and "integrated" projects (projects with multiple types of smart grid devices and applications). The 100 investment grant projects can be divided into two groups: 25 large projects in amounts between $20–200 million, receiving approximately 80% of funding, and 75 small projects of less than $20 million, receiving approximately 20% of funding. One month later in November 2009, the DOE announced funding for about 10 FOA 36 demonstration and storage grants totaling another $600 million.

Starting off with much promise, the DOE projects are proceeding, but the process has suffered bureaucratic delays, questions over tax status, and even some challenges on the pace of funds disbursement. For example, a number of DOE smart grid grant awardees had begun the DOE negotiation process, but soon encountered a potential problem over a question on whether the grants would be counted as taxable income, potentially leading awardees to question whether the grant awards would prove to be too expensive to accept. The DOE announced on March 10, 2010, that the award would be exempt from taxation [30], and soon thereafter, contracts began to be announced.

Federal Activity in 2010

Beyond FOA 58 and 36, a variety of federal initiatives were announced in 2010. A comprehensive federal strategy on carbon capture and storage was announced by the Obama administration in early February, starting with an Interagency Task Force on Carbon Capture and Storage [31]. The 2011 federal budget authorized $40 billion in loan guarantees for innovative clean energy programs, an increase to $302.4 million for the solar energy program, and more than $108 million in new funding to advance and expand research including research on solar energy. These events are but a small sample of the progress seen in 2010. However, with the shift in political power in the 2010 U.S. midterm elections, an era of cost cutting seems about to unfold, leaving the pace of federal activity on clean energy and smart grid an open question in 2011 and beyond.

Renewable Energy Standards (RES) and Renewable Portfolio Standards (RPS)

State legislative activity on smart grid is focused in two areas, renewable energy and energy efficiency. Renewable energy standards (RES) are used to provide utilities with a mid-range target, encouraging them to shift to a more sustainable energy portfolio over time and to encourage economic development and job growth based on clean energy; to date, 30 states and the District of Columbia have such standards [32]. As a matter of policy, the American Wind Energy Association supports the adoption of a national RES [33], stressing the impact on jobs [34]. RES and RPS have proven to be important concepts to drive the growth of the DG market, specifically rooftop solar PV, but many challenge the idea of a national RES, given the widely divergent environments and situations across the United States.

Federal and State Regulatory Influence

The Federal Communications Commission (FCC) [35] regulates interstate and international communications by radio, television, wire, satellite, and cable. Interstate electricity transmission and wholesale electric transactions in interstate commerce are regulated by the Federal Energy Regulatory Commission (FERC) [36]. State regulatory bodies—public utility commissions, public service commissions, and so forth—regulate intrastate electricity activity, specifically investor-owned local distribution companies and, in areas where retail electricity service is now competitive, monopoly transmission and distribution utilities.

The FCC published its *National Broadband Plan* report in June 2010, with Chapter 12 devoted to energy and the environment, underscoring the finding that broadband is essential [37]: "to lead the world in 21st century energy innovation." How so? The report cites four key ways that broadband will enable energy innovation. First, broadband will unleash energy innovation in homes by making energy data readily accessible to consumers, so that with feedback on consumption, consumers can make simple changes, and smart appliances can connect automatically with the grid. Second, different broadband technologies will be combined to modernize the electric grid, making it more reliable and efficient, advancing innovations in renewable power, grid storage, and vehicle electrification. Third, broadband will improve the energy efficiency and environmental impact of the information and communication technology (ICT) sector, specifically our nation's data storage centers and server farms. Finally, broadband will enable a transition to a safer, cleaner, and more efficient transportation sector using real-time traffic information systems and broadband-enabled navigation tools for more efficient route planning and driving for commuters and commercial operators, not to mention more interesting mass transit commuting and transportation substitutes like Web conferencing and telecommuting.

FERC became active in 2010 promoting the incorporation of new types of energy resources onto the grid, with FERC Chairman Wellinghoff expressing particular interest in the incorporation of DR into the wholesale market and the potential of the nascent EV industry to address long-standing issues in the electric industry. In June 2010, FERC released the *National Action Plan on Demand Response* [38] to provide some direction on DR policy. National Association of Regulatory Utility Commissioners (NARUC) [39], a national nonprofit organization, represents the state commissions that regulate electricity, telecommunications, water, and transportation. NARUC holds semiannual meetings throughout the country where commissioners gather to accomplish the work of the association, attending committee meetings, sharing best practices and comparing notes, and meeting with utility, consumer, and vendor representatives to stay informed on industry perspectives. Recent developments indicate that both FERC and NARUC are focusing increasingly more attention to smart grid.

NARUC-FERC Smart Grid Collaborative

In the wake of the enactment of EISA and Title XIII, the NARUC-FERC Smart Grid Collaborative [40] was formed in February 2008 to provide a forum for state and federal regulators to discuss a range of issues to help facilitate the transition to a smart grid, with a special focus on smart grid technologies. Comprised of FERC representatives and 18 state regulatory commissioners, the collaborative has since become a regular feature on the agenda of NARUC meetings. An observer attending successive meetings of this collaborative comes away with a keener understanding of the complexity of formulating new policy in this area, as this large group of intelligent professionals deliberates thorny issues while seated around a series of tables arranged in a big square in a variety of hotel ballrooms every few months, sharing ideas, asking questions, and discussing smart grid in a public forum.

The need to educate policy makers (the average tenure of a state commissioner is just over 3 years) is immense. The need to share viewpoints on a variety of technical, political, and economic issues is compelling, but the challenge to balance the often competing priorities of maintaining and upgrading a reliable grid must certainly be overwhelming to regulators. This deliberative body has become one more tool to help policy makers find their way through the maze.

NARUC and Smart Grid Working Group

At the beginning of September 2010, NARUC [41] announced the formation of a Smart Grid Working Group to be comprised of seven state commissioners representing the diversity of smart grid in the United States, to be cochaired by commissioners from New York and Michigan. This group will help bring focus to NARUC efforts to engage with a plethora of representatives from stakeholder groups, including industry, regulators, and consumers.

Together, the FERC-NARUC Smart Grid Collaborative and the NARUC Smart Grid Working Group represent efforts on the part of state and federal regulators to stay out in front of an issue that threatens to overwhelm them.

Also, the more that the FCC and these two organizations work together, the better, given the chemistry between these two vital technologies, IP networking and electricity, that combine to enable the smart grid. As we look at the rapid adoption of smart grid in more top-down economies like China or more socially progressive regions like Europe, it's hard not to conclude that in the United States we face a daunting challenge by this intersection of a highly complex, high-stakes issue like smart grid and our often cumbersome, disaggregated policy-making processes and institutions. However, time may yet prove the benefit of moving in a more deliberative fashion when it comes to an industry as fundamental to our future as the provisioning of electricity. After all, while the early bird may get the worm, it was the tortoise, not the hare, that won that famous race.

State Smart Grid Dockets

The year 2010 saw progress on many state fronts in implementing smart grid plans of electric utilities. In many cases, the utilities and regulators worked well together to launch projects without major issues. In others, regulators saw themselves repeatedly put in the position of reacting to unintended, negative impacts of smart grid implementations. Viewed together as a trend line, these separate cases help us draw some conclusions.

Oklahoma and OGE

Oklahoma Gas & Electric (OGE) [42], the recipient of $130 million in DOE FOA 58 smart grid investment grants, successfully maneuvered the regulatory process from June to August 2010 and received approval for its Positive Energy Smart Grid program. The OCC preapproved up to $366.4 million in program costs for the system, with the principal focus on smart meters. From June to August 2010, the OCC approval for a large smart grid project shone as a bright light in comparison to the other cases documented in this section.

California and PGE

In September 2009, California utility giant Pacific Gas and Electric Company (PG&E) [43] faced a challenge to its multibillion-dollar AMI deployment, at the time one of the largest and most ambitious rollouts of the new technology in the world, kicking off a trend of consumer backlash to smart grid that has only grown as the year progressed. As if to prove the adage that pioneers are the ones who get arrows in their backs, PGE was sued by consumers in the San Joaquin Valley over abnormally high electric bills that they attributed to their new smart meters. As many insiders suspected all along, the results of an official inquiry revealed in September 2010 that the meters worked perfectly all along, but PG&E had dropped the ball in helping its customers understand the changes underway. CPUC Commissioner Nancy Ryan put it succinctly, "Better communication and customer service will help ensure that consumers see smart meters as something that is done for them, not to them."

Texas and Oncor

Similarly, Texas utility Oncor [44] came under fire for its AMI rollout a few months after PG&E did. With the benefit of going second, Texas regulators reacted much quicker than their California colleagues, hiring Navigant to investigate to identify root causes. Investigation results in July 2010 likewise found the technical performance of the meters impeccable, but those consumers had high bills because of an unusually cold winter and that communication and education could have helped consumers better understand the changes, underscoring the lessons of the California case—consumer awareness will be critical to the success of the smart grid. Public opinion on both these cases appears likely to influence the long-term future of smart grid initiatives and, as the following cases show, the funding for such initiatives as well.

Hawaii and HECO

In Hawaii, there was yet another challenge to smart grid in June 2010, but this time it did not concern a consumer issue per se. The Hawaii Solar Energy Association (HSEA) challenged plans for the pilot, claiming that Hawaii Electric Company (HECO) [45] was "putting the cart before the horse," since the pilot's principal goal was to ratify technology decisions around a smart metering system, but it used a network approach that HSEA claimed would be incapable of supporting future long-term utility needs to integrate applications beyond smart metering, notably, solar PV systems and other forms of renewable energy. When it comes to renewable energy integration, Hawaii is a bellwether state—nearly 90% of its electricity is powered by imported oil, their electricity rates are the highest in the United States by a large margin, and Hawaii also leads the nation with an ambitious 70% renewable energy goal by 2030. The grid will need a major overhaul to accommodate a shift to 30% renewable energy, much less 70%. For now, the pilot is back on track, but Hawaii remains a state to watch as it upgrades its grid.

Maryland and BG&E

Maryland is the "M" of PJM, one of the most congested grids in the nation. Baltimore Gas & Electric (BG&E) [46], the fortunate recipient of $200 million in federal largesse from a DOE ARRA grant, required matching funds to complete the contract and launch its project, for which they would need regulatory approval in a rate case. The trend line emerged ever more clearly in June 2010, as local consumer advocates challenged both the costs and cost recovery mechanisms in the rate case, leading the commission to veto the deal and send the utility back to the drawing board. For a time, it looked like Maryland might be forced to turn its back on $200 million, but cooler heads prevailed and the utility found a way to revise its filing to win regulatory approval in August 2010 and allow federal funds to flow. Two issues became even more apparent in the aftermath. First, consumer groups now had the attention of utilities and regulators and would likely have a seat at the table of any future smart grid hearings nationwide. Second, state regulators would not be expected to blindly follow the lead of the DOE; there would be no automatic state regulatory approval of large smart grid rate cases, with or without federal funds to sweeten the pot.

Illinois and ComEd

In a case that now resembles the proverbial Gordian knot, Illinois utility Commonwealth Edison [47] learned in November 2010 that making it past consumers and special interest groups isn't enough—a utility can stub its toe on procedure, when an Illinois appeals court ruled on a motion by the state attorney general to deny smart grid cost recovery for ComEd. It is not enough to try to reduce costs with innovative programs; ComEd was going to deploy over 100,000 smart meters linked to home energy management systems to lower costs and give customers more control. However, it also matters how a utility seeks cost recovery; ComEd made the mistake of using a special "rider," which carries specific restrictions, which the court deemed inappropriate "single-issue ratemaking." The net-net of this decision will most certainly be to slow smart grid deployments still further and likely to shift more of the burden for installation from utilities to vendors, making it still more difficult for small companies to compete in this industry. It will be hard to challenge other utility managers for being highly methodical and deliberate regarding innovative approaches to industry reform after ComEd's experience and public wrist-slapping.

Colorado and Xcel Energy

One of the earliest, most ballyhooed examples of smart grid innovation though has to have been the Smart Grid City pilot in Boulder, Colorado, where Xcel Energy [48] promised in 2008 to showcase the potential of smart grid with a solution cooked up with a bevy of vendor partners. However, in an outcome described by various parties as pioneer trial and error, miscalculation, hubris, and tragedy, the project suffered from excessive press and challenging delivery conditions, running into multiple project cost overruns along the way (some attributed delays to the decision to lay fiber line through granite in the Rocky Mountains). In November 2010, the utility received preliminary approval for $44.5 million in cost recovery for the project originally budgeted at $15.3 million (March 2008), $27.9 million (May 2009), and then $42.1 million (February 2010). At the time of this writing, the case was under review by Colorado regulators, but based on arguments so far, prognosticators expect the full recovery to be whittled down to a partial recovery [49]. Among the ultimate lessons learned here for utilities may be the value of setting achievable expectations and the wisdom of gaining cost approvals up-front, before proceeding to break ground on a smart grid project.

Victoria, Australia

While not an American state, this state and its similar smart grid experience in 2010 bear mentioning. The leader among Australian states to forge ahead into smart metering, Victoria [50] was first to require its utilities to deploy smart meters to support time-of-use rates. However, in the face of consumer resistance to TOU rates, the state law requiring TOU rate implementation was indefinitely suspended in March 2011, though smart meter deployments continued. Observers expect an expanded perspective on smart grid in Victoria in the next 2 years, as smart meter deployments continue. The Victoria Experience, as some refer to it Down Under, has become yet another object lesson in smart grid in 2010.

State Smart Grid Planning

Whether in an attempt to get out in front of cases presenting few good options, like those cited earlier, or as a simple matter of due diligence and good regulatory practice, state utility commissions opened prospective dockets on smart grid in several states in 2010, most notably the following three cases. These dockets sincerely sought the input of subject matter experts and affected stakeholders in advance of rulemakings and rate cases in order to make better public policy on smart grid. As one commissioner put it, "If it's not in the record, we can't make policy on it."

California Smart Grid Roadmap

In June 2010, the California Public Utility Commission (CPUC) gave final approval to a Smart Grid Roadmap [51], which provides to regulated utilities PG&E, SCE, and SDGE a common model with eight steps to follow going forward as they prepare and implement their smart grid deployment plans: (1) smart grid vision statement; (2) deployment baseline; (3) smart grid strategy; (4) grid security and cyber security strategy; (5) smart grid roadmap; (6) cost estimates; (7) benefits estimates; and (8) metrics. The value of a road map is to balance the need for consistency and interoperability of the system as a whole with the need for individual utilities to tailor their smart grid projects to specific local and regional needs.

New York Smart Grid RFI

In mid-July 2010, New York PSC Chairman Garry Brown issued an RFI [52] requesting comments from traditional utilities, but also from telecoms, software and hardware providers, Internet developers, consumer advocates, and other interested parties to help it develop a smart grid technology road map for New York. The very thorough 15-page document asked about 80 questions in its 10 sections and, in response, garnered more than 50 sets of comments ranging from a single page to three volumes of dense tomes. As a result, NY PSC Docket 10-E-0285 became a great snapshot of the nation's thinking on smart grid in mid-2010. A key similarity in filings from utilities and others was the shift in focus from AMI to grid optimization, where commenters described the potential to start right away to identify and develop projects that make the grid more efficient and lower operating costs.

Oregon Smart Grid Docket

On the heels of smart grid workshops in October 2009, the Oregon Public Utility Commission (PUC) opened a smart grid docket (UM 1460) in December 2009 to develop a 5-year smart grid action plan. To establish this docket, the OR PUC sought help from the Regulatory Assistance Project (RAP), which is financed with support from DOE and other foundational support and staffed by former legislators and regulators, to provide best practices support. As with the NY PSC, the OR PUC seeks to use this docket to bring in insights on smart grid to inform policy making and ensure optimal results as the state moves forward into this difficult area. The RAP connection makes this docket interesting, given that RAP, a source of wisdom in the regulatory community, has taken a leadership role in the formulation of smart grid regulatory policy [53] and that state regulators and staff are likely to optimize these types of proceedings to learn from each other.

Illinois Smart Grid Collaborative

The Illinois Commerce Commission (ICC) established the Illinois Statewide Smart Grid Collaborative (ISSGC) [54] in September 2008 in Docket No. 07-0566 and the collaborative submitted its report to the ICC 2 years later in October 2010. Along the way, numerous workshops provided an opportunity for joint discovery among the multiple stakeholders. The major tasks completed by the ISSGC included: (1) define "smart grid;" (2) understand the range of potential smart grid investments, including potential sources of cost and benefit; (3) identify smart grid policy issues, barriers, and recommendations; (4) define the technical characteristics and requirements for smart grid; (5) develop a cost-benefit framework for evaluating smart grid investment proposals; (6) define utility filing requirements for proposed smart grid investments; and (7) prepare and deliver a final report. The ICC expects to open a smart grid docket in 2011 to continue the work begun by the ISSGC.

Industry Standards and Security

Smart Grid Interoperability Panel (SGIP)

Under EISA 2007, the U.S. federal government gave the National Institute of Standards and Technology (NIST) primary responsibility to coordinate development of a smart grid framework with protocols and model standards for information management to achieve interoperability of smart grid devices and systems.

As directed by EISA 2007, NIST created the Smart Grid Interoperability Panel (SGIP) [55] in 2009 to engage smart grid stakeholders for technical assistance in assessing standards needs and developing the smart grid interoperability framework. In January 2010, NIST issued its first release of a smart grid interoperability framework and road map for its further development, which contains the following key elements: (1) a conceptual reference model to facilitate design of an architecture for the smart grid overall and for its networked domains; (2) an initial set of 75 standards identified as applicable to the smart grid; (3) priorities for additional standards—revised or new—to resolve important gaps; (4) action plans under which designated standard-setting organizations will address these priorities; and (5) an initial smart grid cyber security strategy and associated requirements.

The SGIP organizational structure is characterized by standing committees: (1) the Smart Grid Testing and Certification Committee; (2) the Smart Grid Architecture Committee; and (3) the Smart Grid Cyber Security Working Group. Additionally, five Domain Expert Working Groups (DEWGs) act as experts in certain application areas regarding the requirements of existing and forward-looking smart grid applications. Finally, the technical work of the SGIP is accomplished by applying standards to smart grid use cases under any of the 16 priority action plans (PAPs). More than 1,300 individual members representing over 500 member organizations from 25 countries comprise the SGIP effort.

Smart Grid Architecture Committee (SGAC)

The Smart Grid Architecture Committee (SGAC) has authored and offered up for comment in April 2010 the Smart Grid Conceptual Model [56], a tool for discussing the structure and operation of the power system. The conceptual model defines seven domains (bulk generation [57], transmission [58], distribution [59], customers [60], operations [61], markets [62], and service providers [63]), as well as actors, applications, associations, and interfaces that can be used in the process of defining smart grid information architectures, such as the combined conceptual reference diagram.

Domain Expert Working Groups (DEWGs) and Priority Action Plans (PAPs)

SGIP created five DEWGs to focus industry domain expertise: (1) transmission and distribution; (2) building to grid; (3) industry to grid; (4) home to grid; and (5) business and policy. Consider the Home-to-Grid Domain Expert Working Group (H2G DEWG), for example, which is investigating communications between utilities and home devices to facilitate DR programs that implement energy management. PAPs, on the other hand, are tools used in SGIP to support the analysis and application of standards to the smart grid use cases. The 17 PAP Working Group Management Teams inside SGIP develop PAPs to address either a gap where a standard or standard extension is needed or an overlap where two complementary standards address some common information but are different for the same scope of an application.

Industry Standards Groups

Beyond NIST and SGIP, two groups that bring a multitude of standards bodies together, are a variety of other standards groups more focused on a single set of standards.

U-SNAP Alliance

"U-SNAP" is an acronym for utility smart network access port. The U-SNAP Alliance [64] promotes a universal solution that enables any home area network (HAN) standard to use any vendor's smart meter as a gateway into the home, without needing to add more hardware in the meter. The U-SNAP Alliance promotes a protocol-independent serial interface intended to streamline AMI deployments by promoting interoperability, extending the smart grid directly to energy-aware consumer products.

IEEE P2030 (Smart Grid Interoperability of Energy Technology and IT Operation)

This standards group provides guidelines for smart grid interoperability between the grid and end-use applications and loads. IEEE P2030 [65] created three task forces to deal with power engineering technology, information technology, and communications technology. The P2030 standard addresses interconnection and intrafacing frameworks and strategies with design definitions, which are needed for grid architectural designs and operation.

Open Smart Grid (OpenSG) Subcommittee

The OpenSG Technical Subcommittee [66] was created to foster enhanced functionality, reduce costs, and speed AMI and DR adoption through the development of an open standards-based information/data model, reference design, and interoperability guidelines. Like IEEE P2030, this committee does not create specifications, but recommends developed standards and supply requirements to active standards development groups.

ZigBee Alliance

The ZigBee Alliance [67] mission is to enable reliable, cost-effective, low-power, wirelessly networked, monitoring and control products based on an open global standard to provide the consumer with ultimate flexibility, mobility, and ease of use by building wireless intelligence and capabilities into everyday devices. This standards-based wireless platform is optimized for the unique needs of remote monitoring and control applications, including simplicity, reliability, low cost, and low power. The focus of the alliance includes defining the network, security, and application software layers, providing interoperability and conformance testing specifications, and managing the evolution of the technology. This industry alliance will be instrumental in defining the communication protocols most likely adopted for short-range wireless communication among household devices in the emerging HEMS market, since many smart meters are adopting ZigBee compatibility (e.g., the deployed Smart Energy V1.0 specification with millions of units installed).

HomePlug Powerline Alliance

The mission of the HomePlug Powerline Alliance [68] is to enable and promote interoperable, standards-based home PLC networks and products, ranging from very high-speed technology capable of carrying multiple high-definition AV channels to low-speed, low-cost, low-power consumption PLC for home automation. HomePlug is currently the leading technology candidate for in-home PLC.

Wi-Fi Alliance

As with ZigBee and HomePlug, the mission of the Wi-Fi Alliance [69] is to certify Wi-Fi products (radios based on IEEE 802.11.a, g, n, and so forth), promote Wi-Fi products and markets, and create industry standards and specifications. While Wi-Fi has been late to the game compared to other smart energy systems, its low bit rates, low power consumption, and low cost based on a mature, widely deployed platform make it an attractive technology for next generation smart grid systems. The Wi-Fi Alliance has created a new working group to more fully develop smart grid specifications and recommendations.

Consumer Interest Groups

The emerging smart grid consumers found their voices in 2010. Numerous conferences, executives, and pundits touted the importance of consumers to the smart grid throughout the year, but the point was driven home on Saturday, November 14, 2010, at a 4-hour, open-to-all-interested Critical Consumer Issues Forum, entitled "Focusing on Smart Grid from the Consumer Perspective," jointly sponsored by NASUCA [70] (consumer advocates), NARUC (state regulators), and the Edison Electric Institute (electric utilities), prior to the beginning of the NARUC Annual Meeting in Atlanta, Georgia. Focused in particular on the residential consumer, such constructive dialogue between these three groups was unprecedented and should be taken as a sign of the importance of this issue. An Accenture study [71] released in April 2010, entitled "The New Energy World: The Consumer Perspective," showed that consumer adoption will be a key to the success of utility metering, conservation, and demand response programs, which will require substantial budgets to induce the changes in consumer behavior on which utilities have been counting.

Smart Grid Consumer Collaborative (SGCC)

In March 2010, a group including Best Buy, Control4, Ember, GE, GridWise Alliance, IBM, NREL, and others announced the formation of the Smart Grid Consumer Collaborative (SGCC) [72] to promote the improved understanding of consumer needs in the smart grid universe. The list of sponsoring organizations has grown to include such leading progressive utilities as AEP, Consumers Energy, Duke Energy, Florida Power & Light, OG&E, Progress Energy, San Diego Gas & Electric/Sempra, and Vermont Electric Power and leading industry representatives such as Accenture, Intel, Itron, and OPower.

National Association of State Utility Consumer Advocates (NASUCA)

For more than 30 years, the National Association of State Utility Consumer Advocates (NASUCA) [73] has provided a forum for organizations that represent utility consumers in regulatory and court proceedings. At the time of this writing, membership has grown to 44 consumer advocate organizations from 40 states and the District of Columbia. In 12 states, consumer advocacy is handled by state attorneys general, while in 29 other states, consumer advocacy offices have that role, with directors appointed by governors. Traditionally, NASUCA member organizations have had the role of challenging rate increases in adversarial rate cases. NASUCA filed comments in August 2010 to a DOE RFI on smart grid, focusing their argument on the ability of customers to opt in to advanced rates, the importance of cost-benefit justification of smart grid projects, and the need to educate the widely divergent array of customers on smart grid impacts and issues in tailored programs.

Electric Industry Interest Groups

The U.S. electric utility industry is fragmented, composed of regional utilities formed to address regional needs in this vast, widely diverse country. Domestic transmission and distribution grids (3,121) deliver power to meters (about 140 million) that represent U.S. citizens (over 300 million). Those 3,121 grids break down as follows: 219 *investor-owned* utilities, with larger service territories and most of the major population centers, 2,010 *public power* utilities, with service territories mostly coterminous with city boundaries, and 883 *cooperative* utilities, which are member-owned and were the last to the game, filling in territories that were not served by IOUs and MOUs; and 9 *federal power* agencies. In competitive markets, there are additional retail marketers—in Texas alone, more than 150 retail electric providers (REPs) sell to commercial and residential customers over the existing power distribution networks owned and managed by Oncor, Centerpoint, and American Electric Power. The associations in this section represent the industry and these different segments.

Electric Power Research Institute (EPRI)

The Electric Power Research Institute (EPRI) [74] has taken an active role in fostering a collaborative environment among the nation's utilities and other interested parties to support smart grid research projects and large-scale demonstrations to ready supporting technologies for commercial operation as the smart grid develops. The EPRI Smart Grid Demonstration Initiative is a 5-year collaborative research effort focused on the design, implementation, and assessment of field demonstrations to address prevalent challenges related to integrating distributed energy resources in grid and market operations to create a virtual power plant (VPP).

EPRI Inverter Program

In 2009, the EPRI Photovoltaic & Storage Integration Program (P174) [75] began to study ways to help manage a high DER penetration, with one research area specifically focused on the communication aspects of DER. By mid-2009, this research had led to the launch of a broad industry collaborative to identify a common means for smart, communicating inverters to be integrated into utility systems. The DOE, Sandia National Labs, and the Solar Electric Power Association (SEPA) joined with EPRI to help steer the project, which seeks to identify a core set of potential inverter/charger capabilities that could help enable higher penetration levels of DER and enhance the value of grid-tied PV and storage devices. The project has identified seven smart inverter functional areas: (1) grid connect/disconnect; (2) power output adjustment; (3) power factor adjustment (includes volt/VAR management); (4) storage management (charging/discharging); (5) event/history logging; (6) status reporting/reading; and (7) time adjustment.

TechNet

A national, bipartisan network of CEOs, TechNet [76] promotes the growth of technology industries and the economy through long-term relationships between tech leaders and policy makers and public advocacy of a targeted policy agenda. Together, TechNet's member companies represent more than 1 million employees in IT, biotech, e-commerce, and finance. From its January 2010 release of survey results validating the impact of technology as a tool to make consumers aware of their energy consumption, to a variety of press releases, panels, conferences, and direct lobbying in state and federal governmental bodies, TechNet continued as a voice in support of smart grid and supporting industries and their potential as tools to help renovate the electric industry in 2010.

GridWise Alliance (GWA)

As described at the beginning of this chapter, the GridWise Alliance (GWA) [77] coordinates smart grid organizations, principally by facilitating activities between stakeholder groups and by developing a sound foundation of educational and policy materials and acting as the go-to industry representative for government policy makers and the press when it comes to all issues associated with smart grid. The GWA accomplishes its mission through work groups. The Implementation Work Group focuses on smart grid case studies and value streams.

GridWise Architecture Council (GWAC)

Neither a design team nor a standards-making body, the GridWise Architecture Council (GWAC) [78] is a group of industry leaders, formed at the direction of the DOE, to help shape the architecture of the emerging smart grid. Their principal focus is to provide guidelines for industry interaction and for interoperability between technologies and systems. The GWAC seeks to identify areas for standardization that will stimulate interoperation between the components of the smart grid.

Utilimetrics (AMI)

The original Automated Meter Reading Association changed its name to Utilimetrics [79] to reflect the shift to a broader focus on infrastructure and expanded functionality embodied in the transition to advanced meter infrastructure

(AMI). This industry has provided early leadership in the smart grid space; it is noteworthy that many still think of smart meters when they hear the term "smart grid." At the time of this writing, the AMI industry could be said to have had a very good year in 2010, with 43 of the 50 U.S. states having plans to install smart meters, 20 million deployed so far, and up to 65 million new smart meter deployments planned by 2020, or about half of all U.S. homes, according to Lisa Woods at the Institute for Electric Efficiency [80]. There is still a long way to go, but progress has been significant. Today, virtually all metering companies have acknowledged the importance of the transition to smart meters and a networked infrastructure, and most states have programs underway.

Demand Response Coordinating Committee (DRCC) and Demand Response Smart Grid Coalition (DRSG)

The Demand Response Coordinating Committee (DRCC) [81] and the Demand Response Smart Grid Coalition (DRSG) [82] kept DR issues front and center in 2010, reflecting a growing attention to developing the potential of DR as an energy resource. In March 2010, Pike Research [83] predicted rapid growth for the DR industry starting in 2013, going as high as $8.2 billion by 2020. In June 2010, FERC staff published the National Action Plan on Demand Response [84], which called for a broad-based coalition to implement the plan's three main objectives: (1) technical assistance to states to help them implement DR programs; (2) the formation of a national communications program; and (3) the development of tools and materials for customers, states, and DR providers.

Home energy management systems (HEMS), an emerging new industry that will be instrumental to the success of DR, gained traction in 2010, with a multitude of new market entrants. Two companies highlighted the emerging potential of HEMS in 2010. Intel's entry with a device reference design for ODMs showcased its strategy to foster a new market and new class of HEMS devices using its chips [85]. OPower gained steady traction with entry into more than 1 million U.S. households using a low-tech HEMS solution in the energy bills of 25 utilities, including six of the 10 largest, featuring reports that compare energy consumption between neighbors [86].

Solar Energy Industry Association (SEIA)

The Solar Energy Industry Association (SEIA) [87] represents the solar industry, which saw significant events develop in 2010. A Boston Consulting Group report from November 2010 [88] documented the relentless march towards solar energy grid parity, as solar energy approaches becoming cost competitive with fossil fuels, predicting the gradual fall of the levelized cost of energy (LCOE) for solar PV by half in 5 years, from $0.22–0.26 in 2010 to $0.11–0.13 in

2015, and to as low as $0.09–0.10 or less by 2020. However, the same report also highlighted the challenges that intermittent solar PV will cause the grid as market penetration increases. Under the DOE Solar Energy Grid Integration System (SEGIS) project [89], five teams made progress on projects designed to develop the new products and technologies to enable grid integration of high penetration PV (HPPV) systems.

American Wind Energy Association (AWEA)

The American Wind Energy Association (AWEA) [90] focuses primarily on large remote wind systems. Given that this book has discussed the smart grid from the distribution grid perspective, wind energy has limited impact on the smart grid, since most wind energy flows onto the transmission grid. The same BCG report found on-shore wind energy already at rough grid parity, but that integration with the grid will become an ever-growing challenge, given the remote location of wind energy sites and limited transmission capacity. Still, in 2010, a demonstration project for the Center for the Commercialization of Electricity Technologies (CCET), which received $13.5 million DOE ARRA FOA 36 funding, made significant progress in its plan to tie together wind energy with offsetting DER using smart grid technology. In a unique combination, the project integrates intermittent wind energy resources on one end of the grid using synchrophasors with off-setting balancing resources at the other end of the grid, in a Houston suburb, where a virtual power plant—a combination of distributed solar PV, energy storage, and HEMS—will mitigate the otherwise disruptive impacts of intermittency from wind turbines.

Plug In America, the Electrification Coalition, and the Intelligent Transportation Society of America (ITS America)

Plug In America [91], the Electrification Coalition [92], and the Intelligent Transportation Society of America (ITS America) [93] promote the development of the budding EV industry, hoping to expand the acceptance of electric vehicles beyond EV enthusiasts and early adopters. Historically, Plug In America has advocated on behalf of the development of plug-in hybrids, battery EVs, and other vehicles that use electricity. The Electrification Coalition recently issued reports to educate policy makers and utilities on the potential of EV integration, including the *Electrification Roadmap* and two new reports in 2010 focused more on EV charging infrastructure [94].

Energy Storage Association (ESA)

The Energy Storage Association (ESA) [95], an international trade association promoting the development and commercialization of energy storage technologies, has been hard at its task for nearly two decades. Finally, it seems their message is getting through, as energy storage is arguably one of the most sexy energy topics to date. Highly complex and varied, energy storage technologies have a significant challenge to overcome, namely, the widespread misperception that energy cannot be stored economically. The entire electric industry is organized around this truism, with storage options confined to pumped hydro and a few pilots here and there of newer technologies. Recent technological progress has chipped away at that attitude, as economically viable utility-scale storage solutions have become increasingly available. Energy storage took a big leap forward with the enactment of an energy storage bill, AB 2514, by the California legislature in October 2010 [96].

Edison Electric Institute (EEI)

As the association representing investor-owned utilities, the Edison Electric Institute (EEI) [97] provides smart grid online resources, reports, workshops, and focus at its conferences, roundtables and seminars. At the EEI Roundtable in October 2010, EEI presented a Smart Grid Scenario Project Update [98], showcasing results of its two workshops in 2010 on smart grid, held in Washington, D.C., and Los Angeles. The key takeaways from the workshops on potential scenarios included the following: (1) that technology will have a transformative impact on the utility industry; (2) new market entrants will be strategically positioned between customers and their utilities, leading to customer disintermediation (attractive energy packages from vendors will bypass and displace the utility, leading customers to become less reliant on utilities); (3) a new customer culture (think iPhone) will exist, where real-time information becomes increasingly available via multiple technology platforms; (4) the disruptive impact of smart technologies will occur to incumbent utilities as with telecom companies, regardless of any attempts at regulatory protection; (5) a great many more customers will become less dependent upon utilities; (6) supply resources will come increasingly from both traditional central station systems and distributed resources; and (7) there will be multiple utility business models with varying probabilities of success depending on regional market structure.

EEI counterparts for other utility market segments include the American Public Power Association (APPA) [99], which represents the U.S. city-owned electric utilities, most of which are small distribution-only operations, and the National Rural Electric Cooperative Association (NRECA) [100], which represents the U.S. member-owned electric cooperatives.

Utilities Telecom Council (UTC)

Having represented utility telecommunications issues and utilities since 1948, the Utilities Telecom Council (UTC) [101] has the perspective needed to contribute to the debate on a smart grid transition. From its Smart Grid Policy Summit in Washington, D.C., in April 2010, to its online *UTC Insights*, the smart grid focus of UTC blends technology and policy expertise. In September 2010, UTC released a report sponsored by Verizon [102], titled *A Study of Utility Communications Needs: Key Factors That Impact Utility Communications Networks*, which addressed a key challenge for utilities: deciding when it is appropriate to build and own a network and when it is better to subscribe to services delivered over a carrier network. The design of such hybrid networks will go a long way to the success of a smart grid project.

National Rural Telecommunications Council (NRTC)

Supporting both electricity and telecom rural cooperatives, the National Rural Telecommunications Council (NRTC) [103] provides technology and procurement support to facilitate telecom solutions.

Universities and Smart Grid

Carnegie Mellon, Software Engineering Institute, and the Smart Grid Maturity Model (SGMM)

As described previously in this chapter, Carnegie Mellon hosts and develops new iterations of the SGMM [104] as a tool for utilities to plan and track their smart grid development work. The SGMM also serves the utility industry as a way to track collective progress and provide an industry benchmark for tracking a utility's relative progress. More than 60 utilities have completed the SGMM process to identify the stage they are at in their smart grid development efforts. Carnegie Mellon plans to provide an annual report on utility scoring and collective progress to deploy smart grids.

Virginia Tech Center for Energy and the Global Environment and the SGIC

In July 2009, a team led by Virginia Tech, including IEE Power & Engineering Society and EnerNex Corporation, was awarded a $1 million DOE grant to design and launch a Web-based information clearinghouse to house the expected mountain of data that will accumulate from the DOE Smart Grid Demonstration Projects over the next few years [105]. The Smart Grid Information Clearinghouse (SGIC) was launched in September 2010. Besides data from the demonstration projects, Web site content will include use cases, standards, legislation, policy and regulation, various lessons learned and best practices, and assorted R&D. Over the past year, the team has developed the SGIC database and portal in a deliberate, collaborative fashion, assembling an SGIC Advisory

Committee and an SGIC User Group that together read like a Who's Who list for today's smart grid.

University of Texas and the Pecan Street Project

The University of Texas (UT), the flagship higher education institution in Texas and a global leader in academic research, has staked a position in the emerging clean energy field through multiple departments and, in particular, has proven an active supporter of the Pecan Street Project, as documented in Chapter 5 [106]. The Austin Technology Incubator, closely affiliated with UT's McCombs Graduate School of Business, has a seat on the Pecan Street Project nonprofit board of directors, as does UT's Engineering School. More than 25 UT professors participated in the first phase of the Pecan Street Project in 2009, and now UT's Electrical and Computer Engineering (ECE) has received funding through a Integrative Graduate Education and Research Training (IGERT) award to provide 20 associate professors and graduate research assistants to work on the Pecan Street Project over the next few years. IGERT supports research and educational programs at UT that focus on smart grid development.

Portland State University and the Executive Leadership Institute Smart Grid Seminar

Portland State University (PSU) has created a graduate-level cross-disciplinary seminar on smart grid that is becoming a model for similar smart grid programs at other universities [107]. The seminar, titled "Planning the Smart Grid for Sustainable Communities," addresses two principal audiences. The first audience is graduate students in engineering, IT, public administration and policy, urban planning, business, and economics; and the second audience is current and emerging leaders from utilities, IT, public administration, urban, transportation and water resource planning, architecture and design, business, and other fields. According to the course director Jeffrey Hammarlund, an adjunct professor in PSU's Mark O. Hatfield School of Government, a key challenge of a course like this is the interdisciplinary nature of smart grid, which requires drawing expertise from a variety of fields and disciplines.

Florida State University (FSU) and the Center for Advanced Power Systems (CAPS)

To showcase the amazing variety of academic programs with an interest in smart grid, consider FSU's program, quite different from the PSU program [108]. CAPS is devoted to basic and applied research on power systems technology with a goal to train the next generation of power system engineers.

Smart Grid Media and Events

Smart Grid Publications

A variety of publications have appeared over the past few years; most provide daily updates via e-mail. As with the events, this list is but a sample of publications at the time of this writing that helped smart grid professionals stay abreast of current events. Two publications by Modern Markets Intelligence, *Smart Grid Today* [109] and *Restructuring Today* [110], have carved out a place at the top of the list, with readable format and strong reporting (subscription required). A leading alternative that is available at no cost is *Smart Grid News* [111], a pioneer in smart grid reporting and analysis. Energy Central publications include *IntelligentUtility* [112], an online daily that has grown over 3 years into a must read publication and two others, *EnergyBiz* [113] and *RenewableBiz* [114], that bring more focus from the business side of the equation. Other publications come at smart grid from specific perspectives. For the utility viewpoint, *Public Utilities Fortnightly* [115] consistently delivers valuable information, as do *Electric Light and Power* and *Power Grid International* [116], perhaps with a stronger engineering focus, as does the bimonthly *IEE Power and Energy Magazine* [117]. Finally, additional insights on the renewable energy business are available from the daily online e-mail newsletters *Renew Grid* [118] and *Renewable Energy World* [119] and the Web site greentechmedia.com [120]. Finally, these publications are complemented by publications from the myriad industry associations detailed in this chapter.

Smart Grid Events

There is no shortage of smart grid events in the United States and abroad—in any week this past year, you could probably find a smart grid event to attend if you looked hard enough. In fact, it's not uncommon to overhear a conversation at one of these events words to this effect: "those who are benefiting the most from all this interest in smart grids are the ones putting on the events." Perhaps it was ever so—events are often the front-runners in any new trend, as buyers seek to become educated and sellers look for those early buyers. Policy formation is helped as well by events that seek to educate on the basic facts and bring people together to stimulate dialogue and policy formation. Among the outstanding events that have gained a significant foothold in the emerging smart grid space are three by Clasma Events [121]—Grid Week, Connectivity Week, and Grid Interop; the GridWise Global Forum; Distributech [122]; and annual meetings of UTC, EEI, NARUC, NRECA, APPA, and DRSGC.

Environmental Interest Groups

Environmental Defense Fund (EDF)

Formed in 1967, the Environmental Defense Fund (EDF) [123] has long been an outspoken advocate for the environment. Devoting significant resources and leadership to the Pecan Street Project in 2009, EDF has developed a strategic interest in electric utility smart grid projects and is promoting clean energy best practices to the sector in alignment with their organization's environmental goals.

Conclusion

After this review of smart grid in 2010 on three fronts—national and state level governmental activity, industry standards, and industry stakeholder groups— what do we find in terms of progress for smart grid this past year, and what trends should we watch going forward?

On the national and state level, legislators continue to promote RPS and regulators are moving to gain perspective, educate themselves, and provide necessary leadership within their purview. However, we still lack the federal leadership we need on energy policy, and the U.S. midterm elections in November 2010 did not provide any indication that such leadership will emerge in the near term. Utilities, vendors, investors, and the regulators themselves still struggle with a traditional regulatory construct that at minimum needs a dramatic adjustment to adapt to current changes and to be suitable for the future needs of the economy and society.

With regard to industry standards, we have perhaps the most well-defined progress in 2010. We saw the work at NIST/SGIP, GWAC, EPRI and others unfold deliberately according to a plan as anticipated. The issuance of guiding documents regarding interoperability, security, and industry collaboration provides positive signs that industry standards are developing at a healthy pace.

In the third area, stakeholder groups saw a variety of stakeholders meeting, organizing, discussing, and, with varying levels of success, positioning themselves to maintain their footing on the shifting deck that was smart grid in 2010. The risk of stakeholder group organizational efforts at this point in time is that their positions harden and different groups with widely divergent perspectives continue to speak past each other, leading to gridlock and stagnation. We do see projects that are structured to be inclusive and receptive to experimentation with multiple stakeholder perspectives. The Pecan Street Project [124] is organized around a community integration model. Portland is taking an EcoDistrict approach to sustainability [125] that includes smart grid and looks at the community holistically. At NARUC, the Smart Grid Working

Group is poised to balance the perspectives and interests of regulators, utilities, industry, and consumers.

The principal conclusion of this chapter must be that the industry has paused to reevaluate and adjust to a new environment, exchanging a sense of excitement and expectation in January 2010, when the DOE projects had just been announced and expectations were high, for a more sober calculation of costs and benefits, risks, and trade-offs in December 2010 as the year wound down. What changed in 2010 to bring about such a shift? As documented in this chapter, a shift on two fronts has been brewing, evident for those with their ears to the ground. First, the vision and expectations for smart grid via smart meters had limits and problems that have become more apparent, and second, the consumer perspective on smart meter projects would need to be acknowledged and incorporated at some point, and 2010 turned out to be the year that this happened.

Pecan Street Project Phase One (Chapter 5) took a more expansive view of smart grid, looking beyond AMI, and made the need for consumer engagement one of its three principal conclusions in 2009. Xcel Energy's Smart Grid City in Boulder had by 2009 already experienced significant upward cost estimates and challenges in getting customers to change their ways. PG&E's project had been challenged by consumers in the San Joaquin Valley in October 2009, and a search was underway in early 2010 for a consultant to identify root causes. By March 2010, the formation of a Smart Grid Consumer Collaborative was announced at the Distributech show in Tampa.

The principal smart grid story for 2010 would have to be the recognition of a need for a more sober, measured approach to smart grid, with consideration for all stakeholder perspectives and approval for smart grid projects only after a sound, quantifiable business case has proven out an acceptable risk and return.

The second conclusion to be made in reviewing all the activity in 2010 is that the smart grid is here to stay, which supports our assertion in Chapter 1 regarding the inevitability of the advanced smart grid. Multiple forces, both internal (e.g., aging utility workforce, aging infrastructure, infrastructure complexity, need for operational and capital efficiencies) and external (technology advancement, pending climate legislation and/or regulation, decoupling and unbundling, RPS), combine to lead electric utilities to embrace the need for a smart grid infrastructure transformation.

As noted throughout this book, the principal elements of a smart grid transformation in the distribution (mid and low voltage) sector include: (1) transition to a IP communications network, either through gradual overhaul or outright replacement; (2) grid optimization and distribution automation (DA), including SCADA upgrades, distribution feeder automation, and substation automation; (3) advanced meter infrastructure (AMI) and demand response (DR), including advances inside the home and business, such as smart appliances (SA), home automation networks (HAN), and home energy management systems (HEMS); and finally, (4) distributed energy resource (DER) integration, a broad category that includes distributed generation (DG), especially rooftop solar PV, energy storage (ES), and electric vehicles (EV).

We see an industry evolving towards an integration and network focus (Chapter 2), and we even see hints of recognition concerning smart convergence (Chapter 3), creating innovative industries with new bundles that run along a continuum of increasing integration and market value. The combination of DR, storage, and solar PV (DR-PV) is seen in virtual power plants (VPPs) and micro-grids, emerging components of an advanced smart grid that will be described in more detail in Chapter 7. A VPP system consists of interconnected rooftop solar PV systems (or any other form of distributed generation), interconnected HEMS and DR, energy storage, and management/control systems. With a VPP, the utility can substitute an integrated bundle of distributed energy resources to create a new capacity resource that becomes a dispatchable asset equivalent to a traditional peaking power plant (e.g., natural gas peaking power plant). Finally, a micro-grid is comprised of those elements of a VPP, with more robust DER systems added to enable increased reliability and true energy independence. As will be discussed in Chapter 7, a micro-grid can intentionally island itself from the grid when appropriate, for economic reasons (e.g., a request for curtailment by the grid operator) or for reliability reasons (e.g., imminent outage).

The U.S. market for HEMS, solar PV, and, on a more general basis, DER and smart grid is rapidly evolving, with each of these individual elements in early growth stages. While significant growth is anticipated in all the sectors associated with renewable energy and smart grid, sustained growth remains heavily dependent on governmental policies (e.g., investment tax credits, renewable energy standards, and so forth).

While the energy storage market may be the smallest market now and in the near term based on technology challenges and relative immaturity, it will become the most disruptive when technology advances finally bring costs down to an affordable level for mass adoption in the second half of this decade. Still, each of these individual elements will require integration into a new, transformed energy ecosystem, as described in Chapters 1 through 5.

A final conclusion concerns state regulatory processes, which present a challenge to regulators and utilities alike when it comes to smart grid planning and execution. Traditional cost recovery and rate of return regulation provide control parameters on utilities and attempts to balance the interests of stakeholders in an electricity ecosystem made up of regulators, utilities, consumers, and technology providers. Current rules and processes focus mostly on volumetric kilowatt-hours and the supply side, and have not yet fully incorporated demand side resources and the consumer. However, as shown in this chapter, utilities need new motivations and guidance in this highly dynamic economic climate, characterized on one hand by rapid and disruptive technological changes that create new value opportunities, and on the other hand by the need to incorporate consumers into a historically supply-oriented paradigm.

Consider the forward-looking utility that earns an 11% rate of return, but, through a variety of smart grid programs, has invested and become more efficient, requiring fewer resources to provide its customers better service, but now also returning reduced earnings to shareholders. Now that utility faces still another opportunity to invest in new technology to achieve even greater efficiencies, but on top of reducing its revenues, the rules for depreciation are written for industrial equipment, not technology equipment, which further erodes its revenue potential. Utilities need a new form of risk-rewarding, greatest-value standards that tie compensation to performance against benchmarks and reward decisions that favor prudent, more rapid change and stimulate new value creation. The costs and benefits of an ideal business case, for instance, could be used to guide new utility behaviors that reach towards ever better performance.

In some ways, the electric industry sought these types of results from market mechanisms in the drive to deregulate a decade ago, but now, if regulators are an alternative for market discipline, then they need new mechanisms to reward the desired societal outcomes, such as prudent risk taking, grid optimization, and so forth. Such will be the challenge for regulators in the future—to rise to the occasion and lead the industry to address change, growth, and evolution.

Chapter 7 integrates the themes of all six prior chapters and peeks into the future, moving beyond a world that has successfully grappled with today's changes and implemented a second generation smart grid. Having overcome the hurdles of transition, the evolved industry will be asked to address such future challenges as energy roaming. The twenty-first-century advanced smart grid will face future obstacles and potential in a new world where mobile energy is widely adopted, with peer-to-peer energy trading and other new market mechanisms; a new world where prosumers leverage the grid to share their excess power with each other and provide ancillary services to the utilities to make the grid still more efficient.

Endnotes

[1] Alas, the rapid pace of change in this industry will ensure newer editions of this book sooner than normal.

[2] Apologies to our colleagues at any other utilities inadvertently left off this list.

[3] http://www.cleanedge.com/reports/reports-trends2003.php.

[4] http://www.oe.energy.gov/DocumentsandMedia/EDPR2004_session3_von_dollen.pdf.

[5]   http://www.gridwise.org/.

[6]   http://www-03.ibm.com/press/us/en/pressrelease/28838.wss.

[7]   http://www.apqc.org/smartgrid.

[8]   http://www.smartgridinformation.info/pdf/2961_doc_1.pdf.

[9]   Admittedly, since most utilities building smart grids are not in market competition, char-
       acterizing the trend as a race is really just a metaphor.

[10]   http://www.recovery.gov/Pages/default.aspx.

[11]   http://www.scientificamerican.com/article.cfm?id=wrangling-renewables-and-the-smart-
       grid&page=3.

[12]   http://www.newsweek.com/blogs/the-gaggle/2010/10/18/locke-goverment-moving-fast-
       on-smart-grid-technology.html.

[13]   http://www.nytimes.com/cwire/2010/05/25/25climatewire-anxiety-builds-among-
       utilities-over-the-comm-59064.html?pagewanted=1.

[14]   http://www.nist.gov/index.html.

[15]   http://www.nist.gov/smartgrid/.

[16]   http://www.nist.gov/smartgrid/upload/nistir-7628_total.pdf.

[17]   http://www.gridwiseac.org/.

[18]   http://www.whitehouse.gov/administration/eop/ostp.

[19]   http://www.whitehouse.gov/sites/default/files/microsites/ostp/pcast-energy-tech-report.
       pdf.

[20]   http://www.oe.energy.gov/DocumentsandMedia/EISA_Title_XIII_Smart_Grid.pdf.

[21]   http://www.energy.gov/organization/labs-techcenters.htm.

[22]   http://www.netl.doe.gov/about/index.html.

[23]   http://www.nrel.gov/.

[24]   http://www.lanl.gov/.

[25]   http://www.sandia.gov/.

[26]   http://www.lbl.gov/.

[27]   http://www.pnl.gov/.

[28]   http://www.sgiclearinghouse.org/.

[29]   http://www.SmartGrid.gov.

[30]   http://earth2tech.com/2010/03/10/its-official-no-taxes-on-doe%E2%80%99s-smart-
       grid-stimulus-grants/.

[31]   http://fossil.energy.gov/programs/sequestration/ccs_task_force.html.

[32]   http://www.renewableenergyworld.com/rea/news/article/2010/05/where-the-wind-
       blows-and-sun-shines?cmpid=WindNL-Wednesday-May19-2010.

[33]   http://www.awea.org/pubs/factsheets/RES_General.pdf.

[34]   http://www.triplepundit.com/2010/02/why-a-national-res-is-a-jobs-bill/.

[35]   http://www.fcc.gov/.

[36]   http://www.ferc.gov/about/ferc-does.asp.

[37]   http://www.broadband.gov/download-plan/.

[38]   http://www.ferc.gov/legal/staff-reports/06-17-10-demand-response.pdf.

[39]   http://www.naruc.org/about.cfm.

[40]   http://www.naruc.org/News/default.cfm?pr=77&pdf.

[41]   http://www.naruc.org/SmartGrid/.

[42]   http://www.occeweb.com/SmartMeter/SmartMeter.html.

[43]   http://docs.cpuc.ca.gov/PUBLISHED/NEWS_RELEASE/122937.htm.

[44]   http://www.puc.state.tx.us/electric/reports/ams/PUCT-Final-Report_073010.pdf.

[45]   http://www.greentechmedia.com/articles/read/heco-requests-second-pilot-of-sensus-meters/.

[46]   http://www.intelligentutility.com/article/10/08/bge-wins-smart-meter-approval-conditions.

[47]   http://www.kentlaw.edu/student_orgs/jeel/News_4Fall2010.html.

[48]   http://www.smartgridnews.com/artman/publish/Business_Policy_Regulation/SmartGridCity-cost-recovery-decision-delayed-by-objections-3292.html.

[49]   In January 2011, the CPUC issued an informal ruling that put $16.6 million of cost recovery at risk (http://www.energybiz.com/article/11/01/regulators-curb-smartgrid-city-recovery).

[50]   http://www.drsgcoalition.org/news/media/2010-03-30-SGT-DR_Expert_Welcomes_FCC.pdf.

[51]   http://docs.cpuc.ca.gov/PUBLISHED/NEWS_RELEASE/119756.htm.

[52]   http://www3.dps.state.ny.us/W/PSCWeb.nsf/All/10521538810620598525779000481C75?OpenDocument.

[53]   http://www.intelligentutility.com/article/10/06/regulatory-assistance-project-tackles-smart-grid.

[54]   http://www.ilgridplan.org/default.aspx.

[55]   http://www.nist.gov/smartgrid/.

[56]   http://collaborate.nist.gov/twiki-sggrid/pub/SmartGrid/SGIPCommitteeProductsSGAC/Smart_Grid_Conceptual_Model_20100420.pdf.

[57]   http://smartgrid.ieee.org/nist-smartgrid-framework/bulk-generation.

[58]   http://smartgrid.ieee.org/nist-smartgrid-framework/transmission.

[59]   http://smartgrid.ieee.org/nist-smartgrid-framework/distribution.

[60]   http://smartgrid.ieee.org/nist-smartgrid-framework/customer.

[61]   http://smartgrid.ieee.org/nist-smartgrid-framework/operations.

[62]   http://smartgrid.ieee.org/nist-smartgrid-framework/markets.

[63]   http://smartgrid.ieee.org/nist-smartgrid-framework/service-provider.

[64]   http://www.usnap.org/.

[65]   http://grouper.ieee.org/groups/scc21/2030/2030_index.html.

[66]   http://osgug.ucaiug.org/default.aspx.

[67]   http://www.zigbee.org/.

[68]   http://www.homeplug.org/home/.

[69]   http://www.wi-fi.org/.

[70]   http://www.nasuca.org/archive/index.php.

[71]   http://newsroom.accenture.com/news/consumers+reject+lower+energy+use+as+the+answ
        er+to+reducing+reliance+on+fossil+fuels+and+energy+imports.htm.

[72]   http://smartgridcc.org/members.

[73]   http://www.nasuca.org/archive/index.php.

[74]   http://my.epri.com/portal/server.pt?.

[75]   http://mydocs.epri.com/docs/Portfolio/PDF/2010_P174.pdf.

[76]   http://www.technet.org/.

[77]   http://www.gridwise.org/. Other national organizations that promote the clean tech
        industry from a broader perspective include the American Council on Renewable Energy
        (ACORE) (http://www.acore.org/front) and the American Energy Innovation Council
        (AEIC) (http://www.americanenergyinnovation.org/).

[78]   http://www.gridwiseac.org/.

[79]   http://www.utilimetrics.org/home.aspx.

[80]   http://www.elp.com/index/display/article-display/6349313740/articles/utility-
        automation-engineering-td/volume-16/issue-1/feature/building-the-smart-meters-
        business-case.html.

[81]   http://www.demandresponsecommittee.org/.

[82]   http://www.drsgcoalition.org/.

[83]   http://www.pikeresearch.com/research/demand-response.

[84]   http://www.ferc.gov/legal/staff-reports/06-17-10-demand-response.pdf.

[85]   http://edc.intel.com/Applications/Energy-Solutions/Home-Energy-Management/.

[86]   http://www.opower.com/.

[87]   http://www.seia.org/.

[88]   http://www.bcg.com/documents/file65187.pdf.

[89]   http://www1.eere.energy.gov/solar/systems_integration_program.html.

[90]   http://www.awea.org/.

[91]   http://www.pluginamerica.org/.

[92]   http://electrificationcoalition.org/index.php.

[93]   http://www.itsa.org.

[94]   http://www.electrificationcoalition.org/electrification-roadmap.php.

[95]   http://www.electricitystorage.org/ESA/home/.

[96]   http://carebs.org/AB%202514%20Fact%20Sheet%202010-04-2%20FINAL.pdf.

[97]   http://www.eei.org/Pages/default.aspx.

[98]   http://www.eei.org/meetings/Meeting%20Documents/2010-10-20-StrategicIssues-Smart%20Grid%20Scenario%20Project-%20Jahn%20-%20EEI.pdf.

[99]   http://www.publicpower.org/.

[100] http://www.nreca.org/.

[101] http://www.utc.org/utc/about-utc.

[102] http://www.utc.org/utc/utilities-need-improved-communications-networks-improve-energy-independence-and-sustainability.

[103] http://www.nrtc.coop/pub/us/.

[104] http://www.sei.cmu.edu/.

[105] http://www.ceage.vt.edu/node/82.

[106] http://www.ece.utexas.edu/aboutece/research_detail.cfm?id=59.

[107] http://www.pdx.edu/eli/smartgrid/seminar_content.

[108] http://www.caps.fsu.edu/.

[109] http://www.smartgridtoday.com/.

[110] http://www.restructuringtoday.com/.

[111] http://www.smartgridnews.com/.

[112] http://www.intelligentutility.com/.

[113] http://www.energybiz.com/.

[114] http://www.renewablesbiz.com/.

[115] http://www.fortnightly.com/.

[116] http://www.elp.com/index.html.

[117] http://www.ieee-pes.org/publications/ieee-power-energy-magazine.

[118] http://www.renewgridmag.com/page.php?8.

[119] http://www.renewableenergyworld.com/assets/newsletter/.

[120] http://www.greentechmedia.com/channel/gridtech/.

[121] http://www.clasma.com/.

[122] http://www.distributech.com/index.html.

[123] http://www.edf.org/home.cfm.

[124] http://www.pecanstreetproject.org/.

[125] http://www.pdxinstitute.org/index.php/ecodistricts.

7

Fast-Forward to Smart Grid 3.0

A variety of smart grid events, when strung together in the review offered in Chapter 6, create an emerging trend line that serves to confirm and demonstrate many of the advanced smart grid concepts and arguments found in the previous five chapters. For the insightful observer, an expanded vista of emerging advanced smart grid concepts unfolded before our eyes in 2010. Three principal trends noted in Chapter 6 paint the picture: (1) a growing recognition of a more expansive definition of smart grid beyond AMI; (2) an acknowledgment that a better quantified business case will be needed for regulators to ratify utility smart grid decisions; and (3) the emergence of an empowered consumer perspective that will join the smart grid discussion between regulators, industry, and the vendor community going forward.

Introduction

In this chapter, as we wind up this thread of discovery that we call the advanced smart grid, we review the key concepts offered in this book, we reveal the complexities inherent in the advanced smart grid, we show how these concepts and complexities are addressed in novel planning methodologies and architectures, and, finally, we take a longer look out into the next decade and beyond, to see where the advanced smart grid journey will lead us as Smart Grid 3.0 begins to emerge.

In this chapter, our review includes a discussion of the role a *smart grid architecture framework* can play as a how-to book on building the advanced smart grid. Also, a visionary product that we have termed a *smart grid optimization engine* is needed to operate the advanced smart grid using real-time updates from all managed devices correlated with all utility and end-user systems. Thus, we'll explore in detail the methods of planning, designing, and operating advanced smart grids. Once a decision is made to begin the advanced smart grid journey, the utility is on a path of constant, incremental innovation. The advanced smart grid changes things in a fundamental way. Where current business models, for instance, envision episodic events and projects in between periodic, if sporadic rate cases, the future will feature more rapid change that will necessitate more frequent rate cases, or alternative means of financing structural change, such as self-financing programs that promise a new ability to structure payments according to a new financial rhythm.

To conclude our journey, we look at the transition to Smart Grid 3.0, where a vision of a clean, linked future involves such new innovations as peer-to-peer power transactions, roaming of energy, and integration of energy storage. When pervasive IP networks and computing and energy are commingled with abundant information and nonfuel technology distributed generation, new forms of energy trading will become possible, as straightforward in the future as accessing content over the Internet has become today. The future holds the potential for plentiful clean energy managed over sustainable robust networks.

We have termed this ambitious future Smart Grid 3.0. We believe it is a golden age of abundance, where we manage what we have with greater respect for limits and boundaries, but we also enjoy what we have much more, thanks to sustainable networks that eliminate or minimize waste and encourage easy, even effortless transactions. The future Smart Grid 3.0 will come from applying the concepts and principles elaborated in the previous chapters. As these ideas become incorporated into standards and templates, utilities will gain a practical approach to change that balances current short-term needs for reliability and continuity with longer-term needs of rationality and sustainability. Deploying an advanced smart grid, in our vision, is a means to become far more efficient with our natural resources. For example, if we have the potential today to implement an advanced smart grid and, through grid optimization, reduce the total system energy losses, from 50% down to 10%, then that delta of 40% represents conserved natural resources that could be saved for use by future generations. This is the exciting message of the advanced smart grid, in a nutshell: "A grid designed to leverage innovation to accomplish more with less to enable a sustainable future."

As described in Chapter 6, any given week in 2010 featured at least a few smart grid conferences somewhere around the world. At last count, we estimated over 300 such conferences worldwide in 2010. There is no shortage of discussion on smart grids these days, whether it's online, at a conference, at government locations, or in the office. However, for all the talk about fixing the grid, upgrading the grid, adding to the grid—when the discussion begins with the applications or tools that will define the solution (such as an AMI project), as it often does, that discussion puts the cart before the horse. The smart grid architecture and integrated network design, rather than the application, should be paramount in smart grid planning.

Looking Back

The Inevitable Emergence of the Smart Grid

In our review of the principal concepts in the advanced smart grid vision in Chapter 1, we used real-world scenarios to demonstrate the foundational arguments for an advanced smart grid and to showcase to the world a way for adopting these principles. We began our argument in the beginning of Chapter 1 by showing how much of the complexity in early smart grid projects derives from a decision to begin the project at Layer 7 in the OSI stack, the application layer, which requires significant system integration and results in decisions that limit the future potential of the smart grid. Instead, we offered the alternative perspective that a smart grid project must start with a deliberate process to design the smart grid architecture, whose first step is network design, or face the risk of preliminary application decisions imposing the wrong architecture and limiting the network options needed to meet future needs, not to mention raising the complexity, cost and overall risk of the smart grid project. By starting further down in the OSI stack, at Layer 3, the network layer, a utility avoids the need for so much system integration and drives the design and implementation of a network solution that will serve the needs of the unfolding smart grid forward into ever-growing sophistication.

We further elaborated our belief that the emergence of an advanced smart grid is inevitable, given the fundamental position of electricity in our twenty-first-century lifestyles and economy. Our society has become so dependent on electricity that electricity should be inserted into Maslow's Hierarchy of Needs, at its base. The second reason that the advanced smart grid will emerge is that consumers require greater levels of empowerment. The rationale for technological advances is two-fold: first, to make our lives more convenient, and second, to place individuals squarely at the center of decisions that affect their lives. Because technology empowers individuals in the long run, it is inevitable that the march of technology will ensure that the electricity grid we rely upon does so as well, becoming increasingly responsive to the needs of individuals. This isn't to say that everyone must become an energy expert in the twenty-first century; rather, the individual choices and control over electricity will grow over time, with most change accomplished in the background by technology configured according to the "set it and forget it" method.

The Rationale for an Advanced Smart Grid

Three principles highlighted in Chapter 2—security, standardization, and integration—drive the creation of an advanced smart grid. As digitization becomes normative, the challenges associated with implementing security and the vulnerability of the grid will require an advanced smart grid as the most efficient means to ensure implementation of sufficient *security* measures. The *standardization* of digital devices and networks enables the necessary low cost and rapid adoption and supports the development of an advanced smart grid. Finally, the proliferation of digital solutions and their gradual *integration* in networked architectures will bring about the emergence of the advanced smart grid as the logical and most efficient architecture.

The traditional functional silos of electric utilities—generation, transmission, distribution, and retail services—that have proven so helpful in bringing focus to the specialized, complex tasks associated with generating, transmitting, distributing, and billing electricity in the twentieth century, have today become an impediment to implementing the requirements of the more networked world of the twenty-first century. While scale economies and analog control systems led a utility to organize in silos, the need to share information and manage the enterprise more dynamically in the networked environment of the twenty-first century will lead a utility to reorganize its operations as an integrated energy ecosystem. Moving from vertical to horizontal organization will enable utilities to leverage the capabilities of digital control systems over networks and operate more efficiently. However, to undergo such a transformation, utilities will need to overcome organizational inertia and cultural resistance to new ways of running a utility.

Vendors too will need to adjust to this new way of doing business. Not surprisingly, vendors have historically organized to sell their products and solutions into those silos. When it comes to applications and hardware, managers have readily accepted integrated solutions bundled with a supporting network, but building a smart grid this way shortchanges the architecture and the network decisions, as the single application drives architecture and network parameters. A "just good enough" network, which looked good in the application solution procurement, later becomes suboptimal when subsequent applications must be integrated, and with each application bringing a new network, the number of projects multiplies, and the costs of system integration rise with the complexity of the system. Compatibility and interoperability issues stress the original plans, and expensive work-around projects are not uncommon. There is a better way to build a smart grid.

Stepping back to look at the big picture, it becomes clear that when we describe a vision of the smart grid, we are talking about more than adding new applications to solve old problems, making the grid smarter with each new step we take. We are talking instead about implementing a long-term vision that involves a fundamental redesign of the grid to harness the digital revolution and engage new thinking about architecture and network design based on lessons learned from the Internet.

In fact, if we were to start from scratch today to build a power grid, with all that we know now that we didn't know 100 years ago, we would build an *energy Internet* capable of routing power and information in much the way that the Internet routes bits and bytes today. By necessity, the project plan for the smart grid will be incremental and affordable and may take years to implement, but the plan must be informed by such a long-term vision that accommodates a dramatically different set of needs. The reality of the need to bring consumers into the picture through demand response and the need to integrate distributed energy resources begs the question: How will the system be kept in balance with millions of devices integrated into the grid from all points?

Integration will be a huge challenge, bringing to mind still more questions. How will tens of thousands of home energy management systems cycling hundreds of thousands of appliances on and off be enabled in a grid that is currently blind to activities that lie beyond the distribution substation? How will the grid add multiple EV charging stations that appear in a neighborhood over the course of a few months, when the transformer located at the end of a distribution feeder was designed decades ago to manage a static load limit based on the number of houses it served? How will the system accommodate energy storage units when technology matures over this decade and makes energy storage an economical solution? As grid parity approaches, how will bedroom communities with multiple rooftop solar installations feed their excess power back onto the grid while residents are off at work during the day, when the grid is designed for one-way power flow? How will grid managers address grid stability when large amounts of intermittent energy from wind and solar farms are added? After all, we can't continue to depend on adding ever more generation-based solutions to provide the balancing services needed to accommodate intermittency.

Each of these questions requires a more advanced smart grid that must address new problems of increasing complexity. Each of these questions carries the discussion far beyond the relatively less complex prospect of transitioning from analog meters to smart meters in order to address such current problems as the need for interval data to support time of use rates or the more complex project to add sensors and controls within the distribution grid to improve visibility of grid conditions during outages. As necessary as those steps may be, they don't answer the bigger problem that those questions point to: managing a grid that is becoming far more complex and demanding.

The network architecture to support the complexity of this robust future must support and integrate current and emerging domains, and any future domains not yet identified. Current domains start with centralized generation and its automated generation control systems (AGC) that support reliable dispatch.

The second domain involves generation market operations, and the third concerns the system operations of the utility. Systems in the third domain support both transmission [energy management systems and supervisory control and data acquisition (EMS/SCADA)] and distribution systems [geographic information systems (GIS), asset management systems (AMS), outage management systems (OMS), and emerging distribution management systems (DMS)].

The next critical domain is metering, where the currently popular system is advanced meter infrastructure (AMI), which is comprised of a smart meter end device, a wireless communication network, data backhaul network, and a meter data management system (MDMS) back office function, to provide interval consumption data collection and processing for use in revenue metering and bill production, but also to provide such ancillary functionality as outage management and remote turn on/turn off.

The next two domains involve emerging premise-based systems. Demand response systems consist of a remote control unit connected to a wireless network, used to automate load curtailment as an alternative to dispatching additional supply resources. Distributed energy resources (DER) include distributed generation (DG), electric vehicles (EV), and energy storage (ES). Each of these DER elements is included in some combination of metering and submetering, customer portals, in-home devices (IHD), building management systems (BMS), and home energy management systems (HEMS) to support functionality at the ends of distribution feeders.

Smart Convergence

The story of progress in the modern world has been a story of collapsing cycle times for innovation, as technologies enable the sharing of information and progress and bring innovators closer together. Today progress is accelerating, with digital technologies and the Internet making the world's information readily accessible at nearly zero cost and collapsing historical barriers to bring entire industries together. The utility infrastructures that support our modern lives were, for the most part, built 100 years ago, plus or minus a decade or two, and now they are all being digitized. As these different infrastructures experience similar upgrades to make them smart, we are comparing notes; ideas are borrowed from one industry and applied to the next, and similar patterns begin to emerge, which leads to what we call *smart convergence*. As smart convergence itself matures, it leads to the collapse of still more barriers. Just as the Internet has revolutionized the telecom network it runs on, the advanced smart grid will have a similar impact on the electricity grids that we all depend upon and perhaps all the other infrastructure that supports the economy and our modern lifestyles.

Start with a Smart Grid Enterprise Architecture, Integrated IP Network(s), and SOA

In Chapter 4, we used the case study method to detail the processes, organizational issues, and lessons learned from building the first smart grid in Austin, Texas, from 2003 to 2010. A critical success factor in that pioneer endeavor was recognizing the importance of a smart grid enterprise architecture design process that started with customer engagement and service goals and objectives. Focusing on the customer strategy and associated needs allowed the enterprise to appropriately document the necessary processes, which in turn could then be supported by the appropriate underlying infrastructure, data, and application strategies and architecture design. The key lesson learned from that experience was that any other approach would result in some combination of higher costs, greater complexity and risk, and/or diminished ability to optimize the customer interface. This lesson learned came from trial and error, but also out of the internal struggle between IT and OT and the need to manage to a strict budget.

Envisioning and Designing the Energy Internet

The answers to these integration questions lie in the vision of an energy Internet, as described in detail in Chapter 5. The energy Internet is designed to be resilient and robust, sufficient to support not just the demands of integration, but also future needs as well. The challenge that grid owners and managers face today is nothing less than going back to the drawing board to design an infrastructure capable of meeting the needs of the twenty-first century. Once a smart grid architecture design is in place, project planning and management is needed to transition from the current state to the future state while maintaining reliability.

In Chapter 5, we reviewed the exhaustive community brainstorming that took place in the Pecan Street Project to move beyond the current utility paradigm to a new concept for an energy production and delivery system. The resulting Pecan Street Architectural Framework and the other tools, processes, and concepts are instructive for other communities and utilities as they plan their own futures in a highly dynamic environment.

The final lesson learned introduced in Chapter 5 and elaborated upon in Chapter 6 is the need to engage customers. Making the community more aware of electricity production and consumption issues and of potential impacts on economic and environmental outcomes will strengthen the odds for consumer acceptance of energy technology innovations and the adoption of advanced smart grid programs, associated pricing signals, and dynamic rates.

Consumer engagement will require a period of education and internalization of new energy use habits as a prerequisite. Lower acceptance, lower market penetration, and increasing disappointment in the public eye become foregone conclusions when little effort is made to reach out to the community. The shift to an advanced smart grid introduces dramatically more information into the ecosystem, raising the profile of data privacy issues, data storage and management, and the value and importance of effective communication and operations between the utility and its customers. In fact, successful smart grid programs to date have blended opt-in service components with mandatory infrastructure elements to ensure greater customer empowerment and acceptance.

Today's Smart Grid

As described in Chapter 6, recent events have revealed customer concern over costs, control, and value, motivating utilities to learn more about consumer behavior patterns and their potential impact on utility operations. The fundamental ingredient in community engagement is providing information to customers, ranging from actual marketing collateral stuffed in monthly utility bills to more dynamic digital feedback on energy usage. A dynamic feedback loop of usage information from HEMS devices directly impacts the learning process and facilitates a sense of empowerment.

Besides the in-home devices featured in HEMS, utilities will enjoy ever-increasing opportunities to leverage the multiple screens of customer interactivity, including computers and laptops, smart phones, television sets, and tablets. The content available over all these screens will include applications for smart phones and tablets, Web sites, blogs, social networking tools like Twitter and Facebook, and other types of digital applications.

Advanced Smart Grid Complexities

The next two sections describe a variety of new complexities and associated capabilities that will be needed to manage the advanced smart grid. The first section focuses on grid operations, and the second section focuses on market operations.

Grid Operations

Resource Islanding

Resource islanding, or "islanding" for short, is a term used to describe the voluntary or involuntary off-grid functioning of a premise, community, or local area that has the capability to provide power for itself. Learning to master the islanding of power and load is a necessary step in the evolution to understand how to manage dynamic distributed energy resources on the grid.

The potential disruption to grid operations and synchronization caused by islanding poses a vexing challenge to grid operators, who by and large view islanding as more of a bug than a feature. Few entities have the capability for voluntary islanding today, and notwithstanding the disruptive elements of islanding, mastering this capability would be an attractive option were it to be made available.

Let's talk first about *voluntary* islanding, and the challenge of synchronization. Consider a university that has developed a micro-grid capability and then decides that it is to its economic benefit to operate during a peak condition on its own micro-grid, rather than on the utility grid, so it disconnects voluntarily from the utility grid. How is the now islanded university to reconnect onto the utility grid when the peak conditions subside? In some ways, the challenge resembles stepping off a moving train and stepping back on. The synchronization wave inside the current requires a smooth transition, at the point where the waves of the two systems must be synchronized.

Timing, in this case, is vital, as the physics of electricity are immutable. The sine wave in our 60-Hz grid completes a full cycle from the bottom of its curve and back in 17 milliseconds. The synchronization of two separate electrical systems is a complex task to reengage electricity flow at the sine wave level. The larger system determines the reconnection protocol; the university micro-grid will be required to conform to the status on the utility grid at the point of connection. To accomplish reconnection, the two entities first synchronize their concept of time. Then the grid operator defines when the switch is closed and both grids reconnect. The grid operator needs to reengage the system deliberately, either automatically or manually, at the point when the two grids are synchronous, so that the system is not thrown out of balance. To accomplish this, the micro-grid requires an automatic generation controller, just as a central generation resource like a power plant has, and rules for synchronization and reconnection.

The connection and disconnection of wind farms involve a similar challenge. Over time, the expansion of distributed generation will cause similar issues of engaging and disengaging throughout the grid to arise. Integrating significant amounts of distributed generation will multiply the impact of the micro-grid or wind farm scenarios described earlier, as innumerable small resources are added to the grid, frequently and at different spots.

Islanding presents many questions. If a feeder is set up to handle 1,000 kW, can the operator add more than that with an islanding event? What considerations are needed for grid operations planning? The benefits of devising solutions and answers to these questions include enabling the addition of micro-grids, virtual power plants, renewable energy facilities, and distributed generation, prosumer engagement and control, and regional grid balancing during volatile events that disrupt grid harmony. With islanding, load can be reduced for the regional grid to maintain grid harmony and grid operations, while allowing those within the islanded region to maintain normal operations and full power conditions.

During the historic cold snap through much of the United States in February 2011, ERCOT was forced to order rolling blackouts throughout Texas to keep the grid in balance, because power plant operations at more than 80 power plants had been disrupted by the extreme, sustained cold weather, reducing voltage levels on portions of the ERCOT grid. As more power plants went offline during several days of intense cold weather, healthy portions of the grid that were still producing power were ordered by ERCOT to reduce their production to preserve regional grid stability.

If islanding had been available as an option during that event, Austin Energy's operations, one of the healthy portions of the ERCOT territory, would have been able to avoid the rolling blackouts they endured to maintain the stability of the ERCOT grid. Fast-forward to 2020, as utilities like Austin Energy will have developed programs to encourage individual and collective islanding, and likewise, as ERCOT will have developed protocols for islanding, and a similar scenario will unfold differently. Islanding represents an interesting new reliability and economic solution to grid stability events, but also a complexity that will benefit from the new capabilities that the advanced smart grid brings.

Dynamic Modulation

While we talk about grids that operate at 120 or 240 volts, at a 50-Hz or 60-Hz frequency (the United States runs at 60 Hz), we understand that those numbers are really midpoints in a range and that grid management operations are designed to keep the grid within that range. Dynamic modulation is a micro-strategy for fine-tuning the grid in real time to manage grid frequency and voltage levels with sensors for more efficient resource and load operations and enhanced reliability. Tightening, tuning, and toning the grid in this manner are enabled by smart edge devices and algorithms. Dynamic modulation will be a new tool for grid operators to manage the low and high boundaries of the range to bring about smoother operations.

If we think of current operations as an act of manual multitasking akin to managing a fleet of plate spinners out on a football field [1], then the future will require an automated equivalent, because in the future, that spinning act will move to 10 then 100 football fields, and will expand to include innumerable small backyards and parking lots, and the plate spinners will not be professional plate spinners acting according to a well-defined set of rules, but amateurs intent on meeting their own needs, expressing their creativity according to their own whims and desires. If it sounds like chaos, it certainly has the potential to turn out that way. Dynamic modulation and other tools and well-defined rules of the road will be needed to automate the management function to ensure that the myriad distributed generators and dispatchable demand response resources are orchestrated to balance the grid and make optimal use of these new resources. The advanced smart grid will be needed to bring order to such potential chaos.

Predictive Volt/VAR Control (PVVC)

Predictive volt/VAR control is the ability to anticipate the ratio between power (volts) and reactive power (VARs) on the grid to maintain grid balance. A reactive power condition features more VARs than volts, which is corrected by adding volts. Some devices consume more VARs than volts—motors, for instance. The power electronics associated with PVVC need to account for the constant imbalance and rebalancing of the grid. The power factor is the measure of volt/VAR specific to a location where these events impact grid conditions. Advanced power factor is the capability to adjust the ratio between inductive and conductive load. Starting with a perfect power factor of 1.0, the grid becomes less and less efficient as the power factor declines, requiring more voltage to bring the power factor levels up. Low power factor, the result of uncorrected imbalances caused at the load level, requires more expensive grid operations. Improved power factor holds the potential to require less voltage, but before it can be corrected, the power factor must be measured. As homes in the future shift from a load-only state to a combination of loads and resources, they will gain the ability to export both volts and VARs.

PVVC offers grid managers a new, real-time capability to accommodate increasingly complex grid conditions to analyze and control load changes in real time. Such control will focus on capacitor banks, voltage regulators, and load tap changers (LTCs) to manage the volt/VAR relationship. PVVC will be used to flatten the load profile that goes through the feeder—to push the volts and VARs towards equivalence, thereby minimizing line losses and optimizing distribution feeder functionality, with a goal of a power factor of 1.0. In this approach, more detailed system voltage and VAR information to define more exactly the current state and more defined local directives will contribute to a new tool for power factor correction.

The benefits of PVVC include reducing line losses, leading to reduced generation and carbon footprint, achieving an ideal power factor 24/7/365, and flattening the feeder voltage profile. PVVC will act in multiple ways by capacitors injecting VARs, by generation adding voltage, and by DR reducing load.

Predictive FDIR

Like PVVC, predictive FDIR provides a new capability to address grid conditions, this time by using prediction to improve recovery when things go wrong. Fault detection, isolation, and restoration (FDIR) have three components: (1) fault detection (a monitoring event), (2) fault isolation (a control event), and (3) restoration to normal grid conditions (a management event). These steps will one day be taken automatically by digital devices attached to edge devices, according to preprogrammed algorithms. With an advanced smart grid, switches, transformers, and capacitor banks will come with embedded FDIR capability, providing a self-healing component that is lacking today. Currently, when a device fails, the failure may not be discovered until another event occurs and the discovery process detects, isolates, and restores the fault. Predictive FDIR is paramount to manage involuntary islanding and other disruptive events, because it allows a rapid corrective action by the utility to prevent damaging expensive equipment, thus saving tremendous amounts of money.

As with each of the elements in this list, the complexity of the grid is increasing and these new capabilities are needed to maintain grid reliability and operations. The principal benefits of adding predictive FDIR capability will be to reduce outage duration, frequency, and restoration times, typically measured in such indices as SAIDI, SAIFI, and CAIDI. Such a predictive mechanism enables tighter system management and lengthens the life of system assets. The creation of a logged system of events also provides an accounting record, which can lead to improved management practices over time.

Demand Action

Demand response (DR) has become a well-understood term in the electric utility lexicon, as a utility mechanism to control load using such devices as HEMS and smart thermostats. In contrast, demand action is a utility mechanism for the dynamic dispatch of distributed generation (DG) resources. As these new localized resources become available, grid operators will reach out and use demand action to increase the supply of power to benefit the utility's efforts to balance the grid. When DG becomes abundant and widely distributed, a utility will be able to stimulate energy production on demand based on a price signal and offer to buy.

Market Operations

Abundant Information

The condition of abundant information describes a new state, where a plethora of data gathering devices heretofore unavailable are now deployed and busy producing mountains of data that must be stored and protected. In this case, abundant data can be analyzed with data analytics to produce valuable and abundant information to guide improved grid management on both the supply and demand side. Abundant information also implies the open sharing of information that was often located in silos in the past, kept away from computers, people, and devices that could make use of that information.

Utilities will need to share much more information with their partners and producers out on the edge of the grid. Where utilities before have operated as isolated power experts managing a well-defined grid under controlled conditions, generally with limited information, they will shift to operating as cooperative power experts managed a less-defined grid, but with far more information. Abundant information will not just concern much more data available on the load curve, but will also involve DG producers gaining access to the data they need to better understand their own generation equipment—to do predictive maintenance, for example. Utilities will be able to use this new, abundant information to provide new energy services to create new types of value. A leaky feeder may produce altered volt/VAR conditions that require adjustments to get the most from the active and reactive power that the DG produces. The cooperation and integration challenge facing utilities concerns finding ways to open up to selective sharing of information when they have traditionally protected information for privacy and security reasons.

Prosumer Control

The idea of prosumer control is to enable a consumer transitioned to the new role of prosumer, where they produce as well as consume energy, with new control over new DR and DG responsibilities. Figure 7.1 demonstrates the path that a consumer must go through to move from a state of passivity and lack of awareness all the way up to a state of higher evolution, where the consumer is a

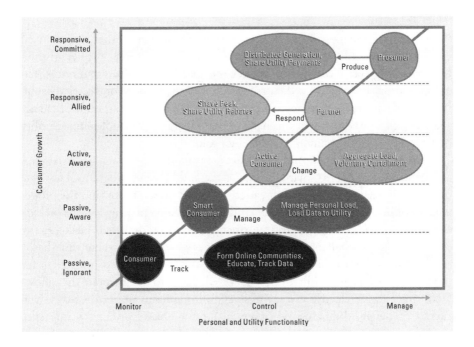

Figure 7.1   Customer engagement development path.

responsive, committed partner of the utility in helping to manage a more rational approach to energy production, distribution, and consumption.

Early utility programs need to provide information to consumers to enable consumption monitoring, so they become more aware of causal relationships. Smart consumers will shift from monitoring to managing their consumption, and along the way provide load data to the utility. The active consumer has become aware, now able to adapt behavior and curtail load on demand. With aggregated behavior and more experience, the customer becomes a utility partner, now responding to utility requests to shave peak and participating in utility rebate programs. Finally, with the integration of consumer-owned DG systems, the consumer becomes a prosumer, enjoying utility payments for edge-based power and ancillary services.

In this transition, consumer engagement represents a tremendous challenge for utilities, which have traditionally seen marketing as relatively one-dimensional, with one-way information distribution via bill stuffers (analog) and Web sites (digital). A growing awareness among utilities as described in Chapter 6 will lead to change, but such change will require extensive resources and considerable time. Most households and businesses will not shift from passive and unaware consumers to responsive and committed prosumers overnight, but when they do, the transition will have an immense impact on the future prospects of an advanced smart grid.

Dynamic Pricing

New dynamic pricing will be reflected in new rate structures that recover fixed costs more equitably, using new, *unbundled* rates that provide greater distinction regarding cost recovery and greater transparency on separate services, enabling a utility to recover fixed costs regardless of the amount of electricity sold. In contrast, *decoupled* rates separate utility revenue from the amount of energy sold, using a rate adjustment mechanism that separates (decouples) fixed utility cost recovery from the amount of product sold.

New rate structure options include: (1) time-based pricing, (2) fixed rates and demand charges, (3) solar customer class, and (4) pilot programs. New rate designs will need to incorporate a variety of rate-design tools to find the proper balance of incentives, subsidies, and equity. Time-based pricing options include time of use (TOU) rates, real-time pricing (RTP), and critical peak pricing (CPP). A central challenge to successful adoption of any of these time-based rate designs will be to elicit the desired customer response and ensure benefit for both the customer and utility; it is a delicate balance to design a rate that allows a customer to see electric bill savings yet still maintains utility profits. Given such a challenge, the perceived potential for time-based pricing to influence energy use should be viewed with some skepticism. Assumptions may not hold true in practice, or may hold true only in specific markets (e.g., large industrial users with the ability to shift operations).

Peer-to-Peer (P2P) Energy Trading

P2P energy trading envisions a world of edge-based energy transactions, where producers of energy on the edge use their grid connections to "ship" power to another party, in an economic transaction where the only service the utility provides is what was called "retail wheeling" in the 1990s. Please see the following Smart Grid 3.0 section for a more complete discussion of P2P energy trading.

Revolutionary Smart Grid Tools: SGAF and SGOE

Moving on from this discussion on smart grid complexities and new capabilities, let us shift our focus now to two new smart grid tools for the design and operation of advanced smart grids. First, the smart grid architecture framework (SGAF) provides grid designers with a handbook of best practices, rules, and methodologies on how to build an advanced smart grid. Next, the smart grid optimization engine (SGOE) is a revolutionary new tool to operate the advanced smart grid.

Smart Grid Architecture Framework (SGAF): A How-To Guide for the Advanced Smart Grid

A smart grid architecture framework (SGAF) is a set of standards, best practices, rules, and methodologies to build a smart grid architecture—answering the how question. As described in Chapter 4, in the discussion on smart grid architecture design, architecture is comprised of the following components: infrastructure (networking, security, computers, and data storage systems), data, applications, and processes. Where the design became the written plan and blueprint to direct activities, the architecture framework (standards, best practices, rules, and methodologies) is the set of instructions on how to build the smart grid. The framework is the cookbook by which one builds a smart grid; the architecture itself is the artifact that describes the smart grid.

Building a smart grid is a new science that requires the application of knowledge and experience. NIST, IBM, and Microsoft have put forward architecture frameworks on how to build a smart grid (and no doubt there are others we're unaware of), but as yet there are not many such how-to guides available in this emerging industry. The set of methodologies in SGAFs encompass a wide variety of disciplines from standards to systems to use (databases, operating systems, programming language, hardware, computers), requirements on storage, disaster recovery, and so forth. Beyond their instructional value, SGAFs are valuable for recording the successes and lessons learned of past efforts and to capture the growing wisdom in the industry.

We need SGAFs because we're entering a new, experimental space. Smart grid is a new science, and like every science, there is a significant amount of art involved. Where does a smart grid designer go for advice on how to build a smart grid? Expertise is found in pockets throughout the industry, and SGAFs are emerging over the past few years, with NIST, IBM, and Microsoft showing leadership. These SGAFs establish the standards, methodologies, and best practices to bring together power, telecommunications, and software and hardware systems to address the needs of an energy ecosystem made up of central generation, transmission and distribution lines, meters and customer systems, and beyond the meter, electric vehicles and charging infrastructure, distributed generation, energy storage, and smart appliances.

An SGAF would help a smart grid designer assemble on average over 100 systems in a rational and effective way to create, distribute, and consume energy more reliably and affordably. Philosophical designs from traditional information technology evolution are transforming the old way of designing the grid, helping smart grid designers to leverage the best lessons learned from past successful transformations in the telecom and computer industries.

A key challenge of the emerging SGAF is that, by default, each electric utility has its own architecture in place. A designer must forge a path to transition from the old architecture to the new architecture in real time. As we have said, the way to build a smart grid will ultimately involve both science and art—techniques and adaptations will work together to craft successful projects. Given the novelty of the industry, few can claim true knowledge of the optimal methodology, much less claim the experience to have gone through this process and come out the other end, so we will all learn from each other, and the evolution of the SGAF will show us the way. The following sections describe the domains and systems associated with an SGAF.

Central Generation

The management of power plants has always been highly automated, as have been the processes and functions around the power plant's output. Unique systems in this domain include automatic generation controllers (AGCs), time series databases, energy hedging systems, wholesale scheduling and settlement systems, generation management systems, distributed control systems, emissions management systems, and laboratory management systems.

Transmission Lines

Transmission system components have become highly automated as well. The most familiar elements are supervisory control and data acquisition (SCADA) and energy management system (EMS). Other prominent systems in this area include load planning systems, asset management systems, and, more recently, emerging synchrophasor systems. This new emerging domain promises a new level of analysis for grid operators, akin to having an X-ray of a point in time that allows detailed data correlation, which will help define a new level of decision making and management of the transmission system. Phasor management units (PMUs) and the software algorithms to correlate the data output will comprise a new synchrophasor system. This new tool has yet to be fully integrated into the transmission management protocols, but promises new powerful levels of management and control.

Distribution Lines

This domain has historically been the focus of the evolution of the smart grid. Distribution infrastructure lacks the full automation that the prior domains (generation and transmission) have enjoyed over the past half-century. As the grid has evolved and the variability and complexity of load has increased, the need to automate distribution lines and associated infrastructure has come to define what we talk about when we discuss smart grid. Key systems in this domain include distribution management systems (DMS), outage management systems (OMS), geospatial information systems (GIS), asset management systems (AMS), vegetation management systems, load planning systems, workforce and workflow management systems, and mobile workforce systems.

Meters and Customers

The genesis of smart metering started with drive-by AMR and a supporting business case to replace pedestrian manual meter readers with automatic readers in vans. Early AMR solutions have been eclipsed by two-way fixed-wireless AMI. The ability to provide Web-delivered services and information came with the advent of the Internet, augmenting the value proposition of AMI. The smart meter remains an evolving trend, as does the way customers receive information on their energy habits. Some unique systems in this domain include meter data management systems (MDMS), customer portals, Web services, call center IVR, online billing, online self-help, conservation management systems, demand response systems, and marketing program and customer loyalty systems.

Beyond the Meter: Home and Office Systems

This domain remains the Wild West of the utility space. Solutions beyond the meter are competitive, as the utility demarcation point terminates at the meter. While many utilities will choose to offer services and/or own infrastructure "beyond-the-meter," many more will choose not to, leaving the door open for a new private industry to emerge. The emergence of the private energy management industry will offer building energy management systems (BEMS) and home energy management systems (HEMS). New and existing companies that will provide these services will include security companies, telecom providers, and cable operators, to name just a few. Key systems that will accelerate dramatically in this space are customer portals, Web services, call center IVR services, billing services, demand response systems, and marketing program and customer loyalty systems. At the time of this writing, a new set of systems was emerging, including EV charging systems, DG management systems, and ES management systems.

As with any new industry, the value of standards is paramount. Evolving standards to guide the creation of these SGAFs range from interoperability to security. The central catalyst for the development of standards comes from NIST, which is empowered by FERC to develop a collaboration between a large number of organizations, including IEEE, IEC, ANSI, SAE, OASIS, and UCA, to name a few. Key standards of note include IEC 61970 and 61968 (application-energy management system interfaces), IEC 61850 (substation automation and protection), IEEE C37.118 (phasor measurement unit communication), DNP3 (substation feeder automation), IEEE 1547 (inverter standard for physical and electric interconnection with DG, EV and ES), IEC 6870 (intercontrol center communications), Open HAN (home area networks), ZigBee Smart Energy Profile 1 & 2 (home energy management communication and data), BACnet (building automation), OpenADR (price response and direct load control), ANSI C12 (metering), OpenGIS and OAGIS (GIS), and, finally, IEC 62351 (information security for power system control). NIST remains the best resource for a complete list of evolving standards in the smart grid industry.

Smart Grid Optimization Engine (SGOE): From Static to Dynamic Grid Operations

The advanced smart grid will require a new way not only to model the behavior of electricity flow on the grid, but to operate the advanced smart grid in real time. Currently, the electricity system is relatively stable and to a degree, predictable. In essence, grid planners regularly conduct modeling sessions, where they model predicted energy flow on the grid based on historic demand and anticipated weather impacts. They produce a network electric model for the week (or whatever planning period is needed) and then reality takes over, as they adjust and manage the grid according to actual load consumption and more immediate data than was available when the plan was devised.

In contrast, the smart grid optimization engine (SGOE) includes functionality that one would find on a network modeling tool, but focuses on dynamic balancing of volt/VAR levels based on real-time data inputs from a multitude of devices. However, the SGOE also provides the ability to control the devices and the grid in real time. How will this differ from what we currently use to manage the grid? First, the SGOE anticipates a much more complex environment, where two-way power flow occurs as the norm rather than the exception.

Two key concepts to keep in mind when discussing SGOE are: (1) updating the system to be able to manage the grid in real time, and (2) managing in a predictive manner, anticipating failures before they occur. As mentioned earlier, we now plan the network electricity flow in advance, and when reality transpires, we use the new data from one period to update the model for the next period. In this way, grid operators tweak the model as needed with each new planning period, but in an ideal scenario, the SGOE would instead optimize grid operations in real time and drive all aspects of grid management automatically, continuously adjusting the model as data came in, not waiting until the next week to issue a new plan.

The next improvement to the current methodology would be to incorporate what-if scenarios to enable predictive modeling to address gaps. To do that, access is needed to all the relevant infrastructure data, such as device procurement date and installation date to calculate the age of the equipment, service status including repairs and reasons, and redeployment, if any. Such asset management data needs to be incorporated because the quality and capability of the assets determine the self-healing potential of the system. The main point of the SGOE is to attain a state of self-healing and automated, efficient operations. Furthermore, the SCADA system needs to provide real-time control capabilities to execute based on defined capabilities and to indicate updates and replacements of relevant assets as needed. We are moving from a current state where maintenance repair crews follow a serial schedule with regular repair cycles, to a dynamic schedule based on actual equipment performance and failure rates, where adjustments and fixes are made just prior to failure or immediately thereafter.

A final benefit from the SGOE is to replace a current inefficient repair and planning system based on stale data with an efficient repair and planning system based on fresh data. The risk of relying on stale data to do repairs, maintenance, and system planning is that the grid becomes overbuilt in places. A "gold-plating" scenario results when stale data leads designers to build the grid for a possible peak event that may never occur. The SGOE provides a much truer picture of grid status that leads to more effective planning and lower capital expenses over time by reducing peak safety building practices via its real-time flattening of the load curve.

Another way to imagine an SGOE is to consider it as a utility ERP; what ERP does for organizations today is fairly well understood and documented (i.e., integrated general ledger plus inventory tracking plus asset management and work orders plus purchasing—all these systems contribute and draw from a common database to be able to synch and maintain a common organizational status view). The SGOE needs to lead a similar transition of the complex electric grid to do what ERP did for the enterprise management: create a more integrated operational model and ensure optimal functioning of the advanced smart grid.

To better understand the potential of an SGOE, let's walk through how an SGOE with true control would work as an engine to promote optimized grid operations. In this scenario, a distribution feeder line is reaching its capacity limit. Currently, there are two principal ways of managing that situation: add more generation to the line (increase voltage) or reduce load (curtailment). An SGOE would expand those options, for instance, by channeling local energy onto the line from a more immediate source, but also by curtailing locally where it was economic and optimal, by coordinating local HEMS and BEMS curtailment contributions. This is a VPP operational model, as introduced earlier in the chapter. Unfortunately, SGOE depends on intelligence at the edge, making this a scenario for the future.

The Smart Grid Journey: From 1.0 to 2.0 to 3.0

The transition to an advanced smart grid perspective, as outlined in this book, may be inevitable, but as the comment on VPP and the SGOE above makes clear, the transition is not necessarily going to happen overnight, nor will it happen without a conscious shift in paradigms. The shift from Smart Grid 1.0 to 2.0 is occurring as outlined in this book and by our theme of simplifying complexity by shifting focus to a more network-oriented perspective. However, the shift from Smart Grid 2.0 to 3.0, which we'll discuss in more detail in the following sections, will be based on a full acceptance of technology and a letting go of the forces that hold us back.

NFTE and FE

To understand the significance of technology in the energy world, let us divide the resources used to produce and manage electricity into two categories based on the use of technology or fuel. On one hand are technology-based resources that do not require fuel. Such nonfuel technology energy (NFTE) on the supply side leverages technology to more efficiently manage the relationship between usable and unusable energy, reflecting the Second Law of Thermodynamics. In other words, NFTE generates power and minimizes the production of unusable energy in the process. Today's NFTE examples include hydropower, wind energy turbines, solar photovoltaic systems, and other forms of solar energy, geothermal exchange and heat pumps, and wave and tidal energy. On the demand side, examples of NFTE include spray-foam insulation, smart appliances, and energy management and control devices and systems (HEMS, BEMS, and so forth).

To produce an abundance of energy in highly efficient ways, Nature has evolved ingenious and harmonious ways to harness the Second Law of Thermodynamics, most notably such processes as respiration and photosynthesis. Nature does this by going small. The key to this value proposition is the emphasis on *optimized, distributed energy production at the micro level.* The system stays in balance by optimal use of energy production by-products in a closed loop to enable corresponding energy-producing processes.

Animals use mitochondria to produce energy at the cellular level with consumption and respiration (i.e., a human body consumes carbohydrates, proteins and fats, water, and oxygen to manufacture energy at the cellular level, producing waste and carbon dioxide as by-products). Plants use chlorophyll to produce energy at the cellular level with photosynthesis, by consuming water, minerals, and carbon dioxide and then using sunlight as a catalyst to transform the water and carbon dioxide into glucose and oxygen (e.g., an apple tree).

In this way, nature provides us with a model of an integrated, distributed network to produce edge power efficiently—a tree is a distributed network with 100,000 micro power plants embedded in its leaves, complete with a distribution system, edge intelligence, and so forth. Similarly, the human body is a distributed network with millions of cells whose mitochondria are micro power plants, managed and controlled by a neural network automates certain functions as an optimal design. Repeating an analogy from Chapter 1, reflexes cause the hand to rapidly jerk back when touching a hot object, without conscious thought, because the nerve pathway becomes more expedient to route a preprogrammed signal to stimulate a response as a survival mechanism. The advanced smart grid will be designed similarly, with routine and urgent responses preprogrammed in algorithms to intelligent edge devices for optimal performance. For such reasons of efficiency, the combination of automated response and distributed intelligence at the edge is the wave of the present and the future.

How energy production is distributed matters greatly in terms of efficiency. Beyond the focus on sustainability described earlier, Nature has crafted a marvelous method of managing risk, widely distributing the means of energy production, so that both production and consumption occurs everywhere (i.e., widely distributed) and energy resources are located near to where they are consumed. This approach calls to mind the way that we have come to manage risk in the financial markets, via a portfolio approach, where a basket of investments has a lower collective risk than a single large investment. Likewise with energy production, nature makes multiple bets and lets it all shake out, with winners rising to the top like cream in a milk bucket. We draw a lesson from these observations, concluding that in this dynamic environment, we face less risk by moving energy production closer to its point of consumption, and by betting on distributed generation over central generation over time.

Energy capture and production at the micro level is already starting to happen and can be seen with any number of emerging NFTE companies, but our favorite is Capstone Metering in Dallas [2], whose revolutionary water meters capture the energy of water flow with a micro-turbine to generate sufficient energy to charge an on-board energy storage device that powers radios in the advance meter with no grid connection.

While fuel-based energy (FE), strictly on the supply side, has been driven by technology advances throughout its history, at its core, FE has always been about the use of fuel to produce electricity. Today's FE examples include fossil fuel–driven coal, gas and diesel (and at some point, hydrogen) electric generators, radioactive fuel–driven nuclear power plants, and electrochemical fuel cells. Energy capture and production using FE has been driven by economies of scale and has become the core of our current electrical grid.

Economics (current lower cost of FE relative to NFTE) and the inertia of the status quo at the technological, economic, and political levels together drive an FE energy policy today; in contrast, NFTE is still viewed as but a complement to our main sources of electricity. The true significance of technology-driven change will become apparent when economies and governments fully embrace an NFTE energy policy, giving it equal or preferred status to the FE energy policy. It may take some time, but such change is inevitable, as is apparent from Figure 7.2.

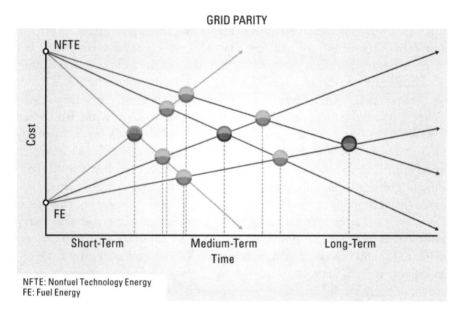

Figure 7.2   Nonfuel technology energy versus fuel-based energy.

The nature of NFTE is to become less expensive over time, driven by innovation and scale production of materials and devices, and distributed electricity delivery through edge-based devices and processes. Innovation and scale production have also driven FE costs lower, and that is likely to continue. However, going forward, the rise of resource constraints and increased demand for FE resources will drive fuel costs and FE costs upwards over time.

It is difficult and argumentative to state the nature of the slope of these curves, but we should be able to agree on the general direction of each as described earlier. Grid parity is a term that defines when the cost of NFTE equals FE. Of the three principal scenarios outlined in Figure 7.2 for NFTE and FE, mapped as rapid change (steepest arrows), medium change, and slow change (gently rising or descending arrows), most of the points of grid parity fall between the short term and medium term, along the two steepest vectors of NFTE and FE. In other words, with rapid change in either NFTE or FE, grid parity becomes more likely sooner than later.

Smart Grid 3.0 Emerges

By the time Smart Grid 3.0 begins to take shape, much will have changed in the utility landscape. Most distribution utilities will have adapted to a proliferation of DER and transformed their grid operations and processes to accommodate hundreds of thousands of new devices, becoming distribution system operators (DSOs) responsible for market clearing transactions and grid coordination. The distribution energy revolution truly begins when grid parity arrives and a new NFTE era is acknowledged and accepted. Utilities that have implemented advanced smart grids will find themselves enjoying the benefits of a flatter load curve and avoided capital expenses for new generation and new distribution system capacity upgrades, which will let them concentrate on new tasks.

As the era of Smart Grid 3.0 dawns, the advanced smart grid will become not just a way to deliver electricity more efficiently, which will bring tremendous value; it will become an entirely new social and transactional platform. New business models, applications, services, and relationships will emerge to leverage new possibilities and new potential created by the shift to pervasive power thanks to NFTE. Just as the Internet has slowly worked its magic to transform our lives at home and at work, the advanced smart grid will provide abundant and affordable power where and when it is needed. Regional and local economic and social success will become based on the adoption of advanced smart grids, because the availability of reliable electricity will become a key differentiator.

In fact, the key challenge for utilities in Smart Grid 3.0 will become reinventing themselves to operate in conditions of abundant supply, as the advanced smart grid unleashes the ongoing, incremental addition of DER devices throughout the grid. In some ways, one is reminded of the transformation in the office environment based on the increase in worker productivity after IT became widespread in the organization. The current dialogue in terms of supply side and demand side will give way to a discussion on quality of power and quality of service, as grid optimization facilitates new electricity market concepts, including energy roaming as EVs proliferate, peer-to-peer energy trading, and energy storage, VPPs, and micro-grids.

A Word on Use Cases

Use cases are helpful to show how an energy ecosystem enabled by an advanced smart grid could look in as little as 10 years, when Smart Grid 3.0 becomes reality. If the concepts outlined in this book are widely implemented, such fictional visions become the logical projection of potential trends and outcomes, a way to integrate this book's many themes and discussions.

Use Case 7.1: Energy Roaming and EVs—Jack and Jill, Up the Hill in Their New EV

In 2015, Jack and Jill love their sophisticated lifestyle in San Diego, one of the most advanced cities in the United States when it comes to electricity. Their host utility, San Diego Gas & Electric (SDG&E), has been at the forefront of research on smart grid for almost 10 years by 2015, and California has consistently led the nation as the most progressive electricity climate.

So, as the price of EVs began to creep down with greater production and market acceptance, and new options became available, Jack and Jill decided it was time to take the plunge and leased an EV. At first, they had been worried about their car running out of juice, known as "range anxiety" in the industry, but their first few months with the new car reassured them. They found that they very naturally adjusted their usage patterns to shift their in-town driving to the EV, and for longer distances they used their second car. Instead of his car and her car, they now had a town car and a country car. Sure, they had to coordinate a little more, but in this way, they both got to enjoy their new toys as well.

SDG&E installed and maintained the car charger in their garage, which offered them three rates of charge. At the fastest charge setting, they could charge their EV in a couple of hours, but they rarely did so because the rates were so high [3]. SDG&E had put them on a special rate for their charger, in exchange for installing and maintaining the charger for free [4]. Technically, the charger still belonged to SDG&E, as the meter on the wall of their house did [5]. The utility leased the charging station to the couple—a nominal leasing fee showed up on their digital bill each month. The couple had entered a new rhythm of charging at the lowest rate overnight, which was more than ample to support the 20-mile commute into town, or even a day of running errands around the house, which could take as many as 40 miles in a day.

While away from the house, they could charge at work, at any of the special charging meters downtown along the street and in parking garages, at the airport, and increasingly in retail locations like the mall, the zoo, and local restaurants [6]. The EV charging/parking spots were always in premium spots, up close, an unexpected perk. Jack and Jill had downloaded an iPhone app for finding available parking spots for EVs soon after they bought the new car [7]. Charging was as easy as plugging in, not unlike putting a gas pump hose into the gas tank. The charging stations had user-friendly interfaces, similar to self-checkout at grocery stores, and an iPhone application was also available to automate the identification process of charging [8]. The high penetration and acceptance rate for solar PV systems in San Diego meant that often when they parked in a garage, their power was coming directly from solar PV panels overhead [9].

After they got more used to their driving patterns and their confidence grew, Jack and Jill had opted into a new SDG&E program, which allowed them to contribute power and something called VARs directly from their car onto the grid (VARs are units of reactive power, a corollary to the volts in our power lines, and are needed on the grid to balance the electricity cycles and keep the grid in harmony). They programmed their car to a minimum charge level, and during the day, their EV would automatically send power (volts and VARs) onto the grid based on economic signals from the utility [10]. Their power bill provided a detailed list of debits and credits to their utility account, not unlike their cell phone bill [11] (except they weren't making money off their cell phone yet). Given that such use could adversely affect the EV's batteries and shorten the life of the EV, the program included a special diagnostic program that monitored battery health and the leasing company had arranged a deal with the utility to compensate it for degradation of their asset, if that were to occur.

A new program at SDG&E linked with other California utilities, so that Jack and Jill could take their car on the road up to Orange County, timing their charge periods to stay within the 100-mile range per each charge. They didn't do that often, but when they did, the new roaming program let them charge up while in OC and have their home account billed at their home rate [12].

Jack and Jill were happy with their EV. They both liked the savings and the carbon reduction that helped the environment. Jill liked the leather seats, the incredibly smooth and quiet ride, and the comfort of the cabin; Jack liked the in-dash GPS with RideTracking, iPhone dock, and integrated Bluetooth and the downloadable "car tones" that project an exterior sound to warn pedestrians of the EV's presence, and he liked to step on the accelerator and feel the torque of the powerful electric motor on occasion, near the house so he wouldn't run dead while out and about. The EV, believe it or not, was quicker than the motorcycle he'd had as a teenager. Who knew?

Energy Roaming and EVs

The explosion of EVs, which will ultimately include new types of vehicles including electric motorcycles and bicycles, will bring the concept of energy roaming into the discussion. Energy roaming calls to mind roaming minutes on a cell phone bill, as used to be the case a few years ago. However, in our case, energy roaming is a positive with payment of a "home" fee regardless of where the energy is consumed, rather than an additional charge on your phone bill.

The possibility of energy roaming could emerge as described in this section, if NFTE becomes more normative. In fact, to put it more strongly, energy roaming as a concept depends on the shift to NFTE as described earlier. As long as our economy is geared to support FE concepts, the payment for energy will remain tied to the fuel. Energy roaming will require bilateral contracts between utilities, and no doubt, some regulatory changes.

Energy roaming will occur inside the utility service territory as a new service, where the key will be to attach a charging event to a particular account. Energy roaming between service territories will be similar to the cellular telecom experience in another way, with a new market created for utility back office accounting transactions. Energy roaming decouples energy service from the commodity energy sale, making it open to third-party sales and services. Energy roaming will entail Internet access to a platform including new applications and content, and new pricing and new service models that depend upon how much energy is consumed (kilowatt-hours), when the energy is consumed (time of day), and where it is consumed (geographic coordinates).

Energy roaming and EVs hold the potential for dramatic impact on the electricity grid. We would go so far as to say that if just half of our vehicles nationwide were to be shifted to EVs, our bulk energy production could remain the same and our energy problems would disappear. Consider an aggregation of parked EVs in Austin, Texas. Approximately 300,000 EVs with 10 kW each of capacity represents 3,000 MW, which is roughly equivalent to the current capacity of centralized power production by Austin Energy. If each EV is driven on average 4 hours each day, it is left parked for 20 hours each day. Assuming that the EVs are charging for 4 hours each day as well, then that leaves 16 hours each day per EV to be part of a distributed, collective energy storage facility, assuming that each were connected and left actionable as a part of an advanced smart grid. On an average trading day, the going rate for a megawatt-hour (MWh) in ERCOT is $60 (on a peak day, that number goes as high as $2,200/ MWh). In that case, 3,000 MWh of wholesale capacity or 3 GWh for a single hour would be $180,000 ($60 × 3,000). Thus, the value of our collective EV storage facility dedicated as a grid asset during its available time, multiplied by

365 days, gives a new annual capacity resource for this utility valued at over $1 billion ($180,000*16*365 or $1,051,200,000).

Use Case 7.2: P2P Energy Trading and Consumer Engagement—Home Energy Management Evolves the Customer Role

6–8 a.m. It's all hustle and bustle as the two kids, Dad, and Mom busily ready themselves for school, work, and a day full of errands and charity work. Thanks to the overnight precooling cycle, the home is cooler than it used to be in the morning, but the family is now used to it—indeed, it's now a great way to start the day. The air conditioner is done for the next several hours—HEMS (see Figure 7.3) has disabled it according to its program—it won't need to start up again until much later in the day [13]. HEMS plugs and breakers attached to CFL and LED lightbulbs in the ceiling and in lamps measure energy usage and ensure optimal usage patterns. Dad scoops up his electronic devices, casually noting that the indicator lights on the HEMS plugs and power strips are glowing red—a good sign, since that means his devices finished charging sometime in the middle of the night and the HEMS devices sensed the change in power draw and shut off power to the transformers, which otherwise would have kept drawing power and wasting electricity—even though the devices were already charged [14]. "So many leaks now plugged by HEMS," Dad thinks and smiles, knowing that those cents have a way of turning into dollars.

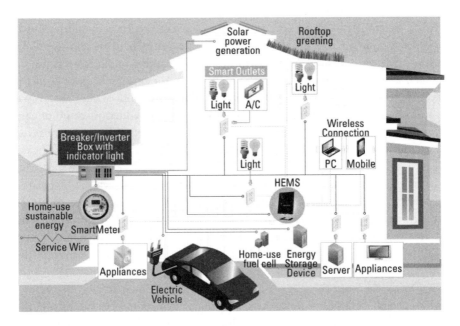

Figure 7.3 Home-to-grid (H2G) design, including vehicle-to-grid (V2G), energy storage-to-grid (ES2G), distributed generation-to-grid (DG2G), and smart devices.

Mom is last to leave, casually glancing at the HEMS unit on the refrigerator and seeing that all is well. In a hurry, she locks the door on her way out. She can't help but note that all the lights are off and the house is eerily silent, as if it had gone to sleep until their return (she doesn't realize that Junior left the LED TV powered up, even though the DVR turned off the picture—it's still drawing power). She smiles, knowing that should she realize she forgot something, she can always access the HEMS personalized dashboard on her smart phone or the GUI interface in her EV [15]. "It's taken some getting used to being so in control," she thinks to herself, "but now conserving energy is more a habit than a conscious activity, and besides, so much is programmed and automated by the HEMS, there really isn't much to think about." As if hearing her thoughts, the HEMS plug sensed the lack of use and the time of day and shut off power to the TV screen, one of the home's biggest energy consumers.

8 a.m.–4 p.m. While the family is away, the HEMS breaker monitoring the circuit on which the family's hot water heater is connected automatically opened the circuit, cutting power to the heater, until the wireless sensor notified the HEMS that it was approaching its minimum temperature and turned the power back on. The HEMS thermostat monitored the air temperature inside their home throughout the day. Thanks to its connection to the HEMS, which is in turn connected to the Internet, the thermostat is aware of events beyond the four walls of the home. For instance, it just learned that a cold front has come through that day, bringing storms and much lower temperatures. Aware of these environmental changes via the Internet and www.weather.com, the HEMS overrode the HEMS thermostat's programmed precooling cycle, which would have started the air conditioner at 2 p.m., similar to its run the night before [16]. And outside, the temperature sensor in the pool failed to register a high enough temperature to make the HEMS breaker engage the pool pump to cycle on to keep the pool temperature at a comfortable, but affordable level— again, because it turned out to be a cool day and the added measure of rain in the pool had kept the water temperature down as well—more energy saved [17].

4–10 p.m. Mom was first home, with Sis and Junior in their new EV and a trunk full of groceries. As the kids took the groceries inside, Mom plugged the car into the HEMS EV charger in the garage. Mom noted that the indicator light turned to yellow as she plugged the car in; sensing the plug in but aware of the time of day, the HEMS EV charger remained off until lower rates would begin after 10 p.m., when it would power on and begin charging to have the EV ready to go in the morning.

With the kids chopping vegetables, Mom sat down in her kitchen office to review the family's power consumption status on her laptop, hitting the HEMS.com dashboard page. The Web site prompted her for user name and password. She quickly reviewed the family's energy status, glancing over her "pictures"—even though she could tell by the green frame around each that all was well, she still enjoyed spending a moment or two to get a fuller understanding—she called to the kids to come see the jar with black jelly beans spilling over—so much carbon offset in only 6 months, well past their initial goals. It felt good to be a part of healing the planet [18]. She noticed that the dial for the solar PV panels was only slowly turning—cloudy weather will do that—usually it was spinning happily away, registering clean kilowatt-hour (KWh) production [19].

One last thing, seeing that they were still well ahead of their goal on their projected electric bill, with 10 days left in the month, she tweaked down the total electricity consumption line to provide a little more of a challenge for her family. Now during the last few days of the month, the HEMS warning light would be glowing yellow, urging the family to go the extra mile to squeeze out the last bit of savings. It really was amazing how 6 short months of data feedback and control devices from HEMS had taught them to change their energy habits and behaviors, while still allowing them all the comfort they wanted. Also, their new detailed consumption data correlated with their total consumption measured by the HEMS—that information came in handy and even had helped them win an argument with their utility 2 months ago, when their bill was out of synch with their HEMS data—a new experience [20]. "Could conserving energy actually be fun?" she mused…time to get back to making dinner, and get the kids to fold the clothes and empty the dishwasher.

Just then, the phone rang—no doubt it was Dad, calling to say he'd be late from work. After the aforesaid notification of a goal reset, he couldn't help but chide his wife for putting the squeeze on the family. Seems that he'd set his personal alarm to notify him of any negative changes in energy consumption, and his wife's goal tweaking had moved the family into a caution zone and triggered his alarm. Relieved that it was nothing to worry about after checking the HEMS dashboard on his work PC, he humored his wife's desire to get a little more out of their HEMS investment. He reminded her that it had already more than paid for itself just last month, 3 months ahead of time. She reminded him of his mother, who lived on a fixed income back in New Jersey. Rates were much higher there, and she suggested they could start shipping their savings to her each month as part of the new peer-to-peer energy trading (P2PET) program that their utility had recently set up. Impressed by her desire to save, and delighted about this new way to share their savings, he found his new conservation behavior rewarding in more ways than one, not nearly as tedious as he'd expected. "What was so good about wasting energy, anyway?" he thought.

Before he got off the phone, he reminded his wife they were due to look into new curtailment rates that had just become available from their utility. They'd have to spend a little time on the HEMS Web site to see how families in other areas with such rates were dealing with the curtailment option.

Next up, he knew, was the potential to join a group of like-minded HEMS users—they'd been tracking the HEMS blog and were looking forward to their first HEMS neighborhood meeting next Wednesday, when they'd all talk about pooling their accounts to get the same rates that large office buildings did. There would also be a presentation by a group that wanted to form an energy MUD (eMUD) in their area. What was that all about? Were they ready for it? He smiled, knowing they were, and knowing HEMS would help them on their new journey to becoming engaged energy prosumers [21].

10 p.m.–6 a.m. As the family sleeps, the air conditioner gradually cools the house according to the cooling algorithms that the HEMS learned in co-ordination with the family (it is summer, and it gets hotter down in the South, even in San Diego). Both refrigerators and the garage freezer are plugged into HEMS plugs, programmed to cycle off and on to provide optimal use of power, but also to skip the chill cycle altogether during a critical peak pricing event [22]. The dishwasher, washer, and dryer all started automatically soon after midnight, on notification from the HEMS that lower off-peak rates had begun (since they got HEMS, the family now loads the machines before bedtime, turns them on, and leaves them, knowing they won't start until told to by the HEMS). Just before the alarm clocks began to go off, the family's pool pump ran one last cycle before shutting down for the day, as instructed by the HEMS breaker attached to its circuit.

Peer-to-Peer Energy Trading

When the combined impact of distributed generation (solar PV), energy storage, and demand response are integrated on the advanced smart grid, and the utility adopts a distribution system operator (DSO) role, then peer-to-peer energy trading will become a new reality. As the name suggests and this short description makes clear, P2P energy trading involves the electric utility only as a distribution resource. The production on one end and the consumption on the other end are conducted absent utility involvement.

P2P energy trading could also occur over longer distances, from two separate electric utility service territories, much as wholesale transactions are conducted today. If electrons truly are innately fungible, why can't individuals and organizations provide their excess power as credits on an accounting ledger, directing the credit to peers who could enjoy the benefits of "free" energy they consume off the grid? Can electricity one day become a more fungible commodity, where the transaction is decoupled from the electron flow? Can electricity be traded the way that airline miles are today, accumulated in an account by a frequent flyer, and then donated to his or her favorite charity at Christmas time? Airline frequent flyer programs have become widely accepted means to build loyalty, and isn't that just what the electric utilities need to be doing?

Trading energy over the Internet the way we currently move content is a matter of accounting, leveraging new grid capabilities brought to the front by the advent of the advanced smart grid. In other words, accounting for inputs and outputs provides a balance with transaction and connection fees. When energy becomes abundant, and systems become more capable, P2P energy trading, or P2PET as in Use Case 7.2, could become a new energy service offered by electric utilities. Consider that we are even likely to see micro-green energy producers emerge as a new small business as P2PET takes hold.

Let's talk in more detail about how P2PET could be implemented at utilities. Energy production at the edge will need to be forecast and planned, perhaps by zip code or neighborhood or, more likely, by distribution feeder level. Power purchase agreements (PPA) between the utility and DG owners are a mechanism to accomplish such planning, because they serve to lock in the capacity sale and tie build or buy decisions on the grid to emerging market pricing and retail needs. At the outset, a utility can use a solar rider or similar mechanism to pay a premium for DG, but as DG becomes ever more common, exceeding centralized generation, then the purchase needs to become more dynamic to more closely resemble the current wholesale market behavior that we understand well. The shift from FE to NFTE ensures that optimal resourcing is enabled on the grid. Load will also need to be aggregated at the edge, feeder by feeder, so that the feeder becomes a pricing node in the local DSO market. Wherever power is produced, depending on the demand and supply on the distribution grid, a dynamic rate can be established, leveraging the smart grid optimization engine to ensure appropriate operational planning and pricing of the resource. The DSO, which may be the utility or a third party, in coordination with the ISO, will need to calculate dynamic pricing by district and by node, down to the smallest possible island, to optimize the production and use it most efficiently at the feeder level. The ISO will focus on large generation and coordination of multiple DSOs for regional system planning. In other cases, the DSO may conduct its own balancing and maintain its commitment to the ISO more independently.

Energy MUDs (eMUDs)

The concept of a municipal utility district (MUD) is well known as a means to provide water and wastewater infrastructure where city systems are not available. State legislatures empower local entities with a publicly elected board the rights to issue bonds, levy taxes, and charge utility rates in order to create infrastructure and deliver valuable services. Over time, some MUDs have expanded their purview to include other services such as garbage collection.

Given the advances to energy technologies as seen in such categories as distributed generation, community energy storage and aggregated demand response, it is not a far stretch to consider the use of an eMUD to provide a local community with new options for energy. As discussed in this chapter, other advances such as resource islanding may likely increase the likelihood of an eMUD emerging in the near term.

Use Case 7.3: Micro-Grids, Integrated Energy Storage, and Packet Power—Building Energy Management System (BEMS) in Action in the Future

12 a.m.–8 a.m. Nighttime is the major off-peak cycle in any utility service territory, when electricity rates are the lowest. Small commercial businesses have a variety of strategies to take advantage of lower electricity prices, and BEMS (see Figure 7.4) enables those strategies, even automates many of them. Precooling, a tremendous energy saver, is one of the best. The air conditioner or, for larger businesses, the chiller, integrated with on-site energy production and storage, precools the office/store/warehouse/worksite according to the cooling algorithms that the BEMS learned in coordination with the business manager in the first month of operation (larger businesses will have a dedicated energy manager). Charging of forklifts and other battery-driven equipment—a category that now includes PEVs—is best accomplished overnight to avoid the high spikes in energy consumption that can prove costly under electricity rates that include a "demand" charge.

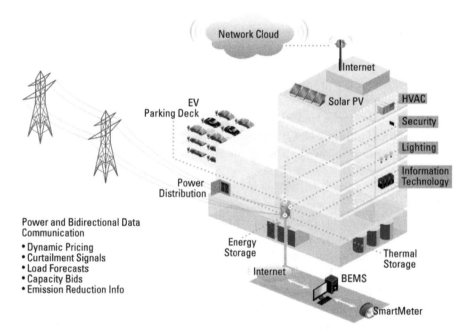

Figure 7.4 Building to grid (B2G) design including vehicle-to-grid (V2G), energy storage-to-grid (ES2G), distributed generation-to-grid (DG2G) and smart devices.

Many small commercial businesses were early adopters of the new EVs when they became available starting in 2012. Thanks to BEMS plugs and power strips, transformers can stay plugged in but no longer draw power after completing their charge (a little-known fact is that transformers used to charge devices that run on DC power continue to draw AC power, even after the devices are fully charged). Thanks to data feedback from the BEMS and its online dashboard, Ms. Small Commercial (SC) decided to have her outdoor lighting changed to compact fluorescent (CFL) and LED lighting systems to lower consumption. Ms. SC is now much more aware of how much electricity each aspect of her business consumes; she even has a complete strategy for managing her businesses energy costs—something that wasn't really feasible before the advent of BEMS and its data feedback cycle [23].

8 a.m.–6 p.m. First to arrive, Ms. SC still opens up shop and goes through her daily routine, but her routine has shifted since she became a BEMS customer and grew more aware of her consumption and, increasingly, of the potential for conservation. She no longer walks through the building, turning on all the lights, and lowering the thermostat setting to get the building cooled off after it warmed up overnight. The building was already precooled according to a predetermined electricity conservation plan and lower overnight electricity rates. BEMS plugs and breakers are programmed to turn lights on and off according to a schedule based on room occupancy—"that's one less chore to start the day," Ms. SC thought, appreciating the extra time she now had.

Instead, she went straight for the PC to log in to her personalized BEMS energy management dashboard and prepare for her weekly staff meeting at 9 a.m. As she began to scan the different screens to track the performance of her business to its preset electricity goals, she couldn't help but recall how much things had changed in the 6 months since she began taking control of her business electricity use. Always on the lookout to improve the bottom line, it seems that she had made a habit out of talking up electricity costs at staff meetings. She always needed to remind her staff to turn off lights and try to save electricity, but nobody ever seemed to have the same sense of urgency she did as a small business owner. Despite her best efforts, it was common in the pre-BEMS days for staff to have a running battle over the room temperature, fiddling with the thermostat despite office rules that clearly outlined office policy and required an energy-saving setting of 78°. Now with BEMS automation protocols and recommended precooling strategies, the office staff has reached a general accommodation and those arguments have become a thing of the past. Who knew saving energy could result in increased workplace harmony [24]?

Back to her work on the BEMS energy management dashboard, Ms. SC was most concerned about office progress to their monthly consumption goals, although with each month's progress, she could see how changed behaviors were, allowing ever tighter goal setting and electricity cost reduction. After just

6 months, they had already dropped their electricity bill by 30%, progress that had eluded her before she had BEMS working in the background on her behalf. She printed several screenshots showing the business energy performance, including the resulting reduction in greenhouse gas emissions. She planned to share those with her staff, reflecting how much more pleasant it was to congratulate them on progress rather than harangue them on the need to save, and she knew that, like her customers, her employees were increasingly more concerned about "green" issues.

She had already incorporated the good news about her new conservation-oriented workplace in the latest marketing collateral—anything to get an edge on the competition—and the new solar panels had just gone up last month, which, with their integrated energy storage unit, promised even more efficient operations. Already, the sun was inching up in the sky and the dial on the computer screen was whirling away, now and again running counterclockwise as the numbers ticked down, not up [25].

A crazy thought passed her mind—one could even say that her business had added a new profit center, given that she was using her rooftop to generate electricity. On a whim, she jumped over to the Web site of that outfit in Dallas that was selling microturbines to fit on the eaves of buildings and capture the upflow of wind to generate electricity. Although still skeptical of how much energy she could actually generate that way, she had become more aware since getting all the data feedback on consumption that every little bit counts when it comes to electricity savings. Not only do little changes add up to big savings, but highly visible conservation—like solar panels and micro-turbines—also send a strong signal to her customers and prospects, even becoming another part of her marketing strategy [26]. With the advent of P2P energy trading, which the utility had introduced as a new program called P2PET, she realized that she could indeed start exporting her savings. She imagined a program where her employees could access free power as an employee incentive. Thanks to BEMS and other advances, she had newfound confidence to make complex electricity decisions about her business.

After the staff meeting—"that went well!" she thought—she had an appointment with a representative from the electric utility to discuss a new rate program for early adopters like herself, small commercial users now experienced with the BEMS system. Acting on her behalf, BEMS had matched her business load profile with other similar businesses and prompted the utility to make an invitation to the group of like-minded users to join a new load consolidation program. By agreeing to work in unison with the other businesses, jointly cooperating with the electric utility to lower peak demand to avoid expensive electricity production or purchase, her business and others in her group would qualify for a new rate class that was the equivalent of a large commercial ratepayer [27].

Thanks to 6 months working with the BEMS dashboard and the equipment in her office, Ms. SC had become well aware of her own load profile, how it compared to national averages for businesses of her type and size, and the impacts of high demand during peak times. Working independently, she had already managed to make the necessary changes in behavior to voluntarily lower her peak consumption and avoid demand charges, but now it looked like she would have a new opportunity to actually save significantly more money for doing what now came naturally to her and her staff. Next week, she would get to meet the other businesses in her new group—they planned to meet once a month to compare notes and best practices, and encourage each other to save even more [28].

As she headed to lunch, she glanced at the BEMS indicator light mounted over the door, reassured that it glowed green. She knew that the adjusted goals would move it to yellow as the end of the month came near, giving her and her staff that much more incentive to be more mindful of consumption in order to meet their monthly goals. At lunch, she compared notes with a friend on the chamber board, who wanted to know more about the BEMS approach. She was surprised at the focus of the conversation, which turned more frequently to the new carbon credit market. She knew that her business had not only lowered its electricity bill, but had also been responsible for eliminating tons of CO_2 in just 6 months of electricity conservation—that was evident from the BEMS dashboard carbon tracker—but she had overlooked the economic potential of trading in new carbon credits. Back at the office, she clicked through and found a BEMS program to manage those credits on her behalf, pooling with other BEMS users to get optimal market value [29]. "How had I managed to overlook that?" she wondered. "But so much was different in just six months, and they had come so far," she reassured herself. She made a note to investigate this new revenue opportunity.

6–10 p.m. First to show up, last to leave, such was the life of the small business owner. Ms. SC took one last look at the BEMS dashboard—after all, she could easily check it at home if there were anything she'd forgotten—then went through her new office shutdown routine. It didn't take nearly as long to shut down the office by clicking through computer screens as it did to walk around the building and inspect light switches, systems, and so forth. No longer strictly reliant on a mental checklist, now she merely had to quickly review dashboard screens that monitored the BEMS, thermostat, breakers, and plugs, detailing current operational status as well as up-to-the-minute energy consumption levels, comparing them to preset goals based on best practices and industry norms. Not only did it take less time to manage her electricity consumption, it was far more effective than the old system of manual checks and balances, individual smart thermostats and other utility efficiency programs. Thanks to BEMS, electric expenses had been transformed from one of the hardest-to-manage line items on her income statement to one of the easiest. In fact, she pondered what life would be like if there were a BEMS system to help her manage all the other items on her list.

Integrated Energy Storage, VPPs and Micro-Grids

Electrical energy storage at utility scale remains both the Holy Grail for electric industry redesign and one of the most vexing energy technology challenges. Storage applications may include: *peak shaving (load leveling) systems* to help commercial and industrial users manage electricity costs under variable utility tariffs and to help utilities manage generating assets to minimize waste; *renewable integration systems* to help power producers, utilities, and end users cope with the inherent variability of wind and solar power, transforming it into firm, dispatchable power, and to better match peak wind and solar output with peak demand; *power quality systems* to protect commercial and industrial users from interruptions that cost an estimated $75–200 billion per year in lost time, lost commerce, and damage to equipment; and *transmission and distribution support systems* to help utilities reduce grid congestion, defer upgrades, and minimize waste.

Because utilities still depend on spinning reserve and other supply-side strategies to ensure reliability—and even do long-term planning on the assumption that energy storage will not be viable in the foreseeable future—adding energy storage into the system will significantly disrupt the current energy ecosystem. Implementing an advanced smart grid will help utilities to develop a vision based on the potential of affordable energy storage, where they will move from R&D to active trials of various technologies in different parts of their service territory.

The use of energy storage does not need to concern any loss of comfort or convenience, but rather to accept some minor sharing of the resource in a collective strategy to pull in a new resource that has never existed before. An advanced smart grid will enable solutions that leverage distributed energy resources, tapping into some that haven't yet been conceived. Integration with other assets, including distributed generation, demand response, EVs, VPPs, micro-grids, and smart grid technologies will be vital to maximizing the value of each individual new resource, but also of the collective advanced smart grid. With a network and smart grid optimization engine, all these distributed elements can be optimized and their potential realized.

Challenges in creating a utility energy storage program from a strategic perspective will include not only energy storage integration, but also designing the system and prioritizing energy storage locations. Energy storage is likely to be located where there is congestion on the grid.

The role of energy storage in a disaster recovery situation will need to be considered. Energy storage will be valuable in restoration of electricity service after a massive outage. Prioritizing distributed generation and energy storage for disaster shelters will buy utilities time, since there will be a minimal amount of power assured in those spots in the case of a prolonged outage. Colocation of DG and ES with disaster shelters supports utility goals in a disaster recovery plan.

Virtual power plants (VPPs) describe a demand-side alternative to accommodate growth in peak demand to the traditional supply-side alternative of adding a natural gas power plant, commonly referred to as a *peaking unit* or a *peaker.*

At the micro level, VPPs will require technology to be refined at the scale of a single distribution feeder or neighborhood.

At the macro level, integrating a VPP system to the grid will require the ability to manage new levels of complexity in remote sensing, control, and dispatch. Integration of the complex VPP systems will require the utility control center to be able to "see" the status and availability of such distributed capacity. The complexity of system dispatch with these types of resources will require automated decision-making to signal a direct load curtailment condition to DR resources to make optimal use of these new resources and integrate into the larger system portfolio.

New control software now available in the market is designed to enable utilities to manage the numerous and complex dispatch requirements of such distributed energy resources as VPPs. Intelligent edge devices such as smart meters, smart routers, and smart inverters are now capable of communicating their operational status, calculating the ramifications of their actions on their surrounding environment, and making decisions to change their state in real time so that the network becomes self-healing and self-adapting. Autonomous, edge-based decision-making maintains safe energy flows, minimizes service disruptions, and, perhaps most importantly, helps to avoid catastrophic damage. Such a distributed network of smart devices connected to VPPs will provide them the intelligence they need to create new demand action capabilities that integrate new edge-based resources seamlessly with the advanced smart grid.

VPPs and micro-grids represent an emerging resource and application of creative bundling of multiple technologies. New concepts and thinking about how a grid works will be needed for these approaches to take hold. For instance, while initial adopters of these solutions may have off-grid operations in mind (Step 1 could be called "disconnect"), the actual realization of these concepts will be difficult. Step 2 could be called "independent operation," and it will borrow from our knowledge of how we operate the grid today and concepts in this book about future operational protocols. Step 3 is likely to concern the previous discussion on resource islanding, with a focus on "reconnection for reliability." Step 4, which we could call "replacement," will finally be realized as NFTE is implemented as policy. As VPPs and micro-grids mature, they will become the LANs and WANs in the Smart Grid 3.0 world, connected by an energy Internet and offering new levels of independence and interdependence that we have barely contemplated today.

The twenty-first century demands a new set of organizing principles when it comes to electricity. The changes will be on multiple dimensions, as shown in Figure 7.5, starting at noon (using a clock metaphor) and moving clockwise around this circle of progress of the twentieth-century grid fundamentals and the sea change to a new set of twenty-first-century grid-guiding principles.

- Where the early grid developers embraced access to cheap, plentiful fossil fuels like coal, petroleum, and natural gas, building ever larger, more efficient coal-fired power plants, today we're challenged to avoid the carbon that results from burning fossil fuels.

- Where grid resources on the supply side expanded to meet a growing population and increasing use of electricity, today we're challenged to involve consumers to avoid peak load by making better use of existing and new types of distributed energy resources.

- Where production and distribution of electricity were central to planning and building the grid, today we also focus on our built infrastructure to use energy more efficiently and to better understand energy consumption behavior patterns.

- Using cost plus ratemaking to establish revenue targets for regulated utilities, the keystone of the twentieth-century "regulatory compact" that allowed monopoly franchises is giving way today to considerations of ecosystem impacts, decoupling, and more frequent true ups to meet the needs of society and a variety of stakeholders in a more dynamic environment.

- When information used to be scarce and expensive, utilities devised ingenious, artful methods to plan and operate the complex grid, but in today's environment of abundant, low-cost information, we are challenged to add sensors to gather ever more data, and then use data analytics to plan, operate, and manage complexity.

- Late in the twentieth century, retail competition showed faltering progress and potential, and while markets still have a prominent role to play, in the twenty-first century communities out on the edge will leverage social networking, smart mobile devices, and other lessons learned from the Internet to identify and deliver greater value.

- Finally, where AMR evolved into AMI and led the charge to deliver a smart grid, in the twenty-first century we will need a more expansive

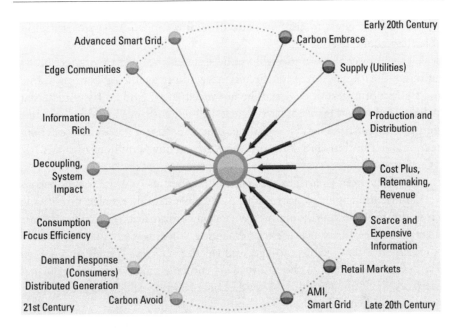

Figure 7.5 Electricity economy transition.

definition of smart grid, based on a smart grid architecture to incorporate edge power by design, which we call the advanced smart grid.

The Advanced Smart Grid: Edge Power Driving Sustainability

The advanced smart grid offers the utility industry and related stakeholders a new framework and operational paradigm for energy creation, distribution, and consumption, but also a transformational social and transactional platform.

Our work and play—our very lives—will be transformed as the advanced smart grid becomes pervasive. It has been said that the smart grid will exceed the impact of the Internet, and we would not dispute that assertion. The advanced smart grid will accelerate job creation, and, more importantly, it will stimulate the emergence of edge power and foster a new age of nonfuel technology energy (NFTE). The advanced smart grid will enable end-to-end cyber security from the device through the network to the core of the utility and back.

The advanced smart grid will enable a transition from the incredible complexity of today's grids to an enhanced simplicity of use by the utility. The advanced smart grid will be built with an open standards-based smart grid architecture framework to support data dissemination on TVs, smart phones, tablets, and computers.

The advanced smart grid will not be about any single network technology but rather the integration of multiple IP networking technologies using a single smart grid optimization engine. The advanced smart grid will stimulate data analytics and new transactions to engage utilities, customers, and independent energy producers in a new, self-healing, interactive energy ecosystem of energy creation, distribution, and consumption reaching out to millions of smart edge devices.

The advanced smart grid design, as described in Figure 7.6, will become the manifestation of a new horizontal energy ecosystem that replaces utility silos across the four existing domains: generation, transmission, distribution, and metering services, as well as the four emerging domains: demand response, distributed generation, energy storage, and electric vehicles. The advanced smart grid design will integrate the utility with end consumers in a variety of new interfaces, including building-to-grid, home-to-grid, vehicle-to-grid, energy-storage-to-grid, and distributed-generation-to-grid.

The advanced smart grid will introduce us to new concepts and capabilities and new ways of thinking about energy and how it impacts our lives. The advanced smart grid will enable a new social and transactional platform for clean and abundant power.

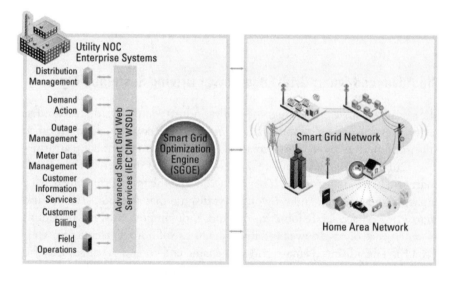

Figure 7.6   Advanced smart grid design.

Endnotes

[1]   The vaudeville act on *The Ed Sullivan Show*, where a professional plate spinner lined up a series of pool cues and proceeds to keep increasingly more plates spinning on them, is an excellent example of what grid operators will have to contend with when DER blossoms; see it on YouTube at http://www.youtube.com/watch?v=Zhoos1oY404.

[2]   Capstone Metering, LLC http://capstonemetering.com/.

[3]   Special charging rates discourage rapid charging that stresses the utility system.

[4]   Special charging rates encourage usage during nonpeak hours, when renewable wind energy is abundant.

[5]   Smart EV charging system is a new smart end device equipped with intelligence.

[6]   EV charging becomes a new revenue source for utilities, which control location to accommodate grid acceptance of EV charging.

[7]   Incentives guide drivers to charging location sections that are conducive to a healthy grid.

[8]   Account correlation with mobile charging enabled by advanced smart grid.

[9]   Combination of solar PV and EV enabled by smart inverters.

[10] Predictive volt VAR program and two-way power flow use EVs to help with balancing.

[11] Digital billing draws usage information from a common database.

[12] Two-way information flow enables "mobile" rates that are driven by usage and account, irrespective of location.

[13] Prechilling the home is an example of thermal energy storage as a DR strategy.

[14] A good example of a strategy to address the challenge of "vampire power."

[15] Multiple screens will provide access to HEMS information and remote control for ultimate ease of use and optimal fine tuning and control.

[16] A new, more expansive definition of a "smart" thermostat.

[17] Pool pumps consume more energy than almost any other appliance in the home.

[18] Tying conservation goals and motivations such as climate change to data feedback is a powerful means of making changes in energy consumption behaviors to be permanent adjustments and new habits.

[19] Information on distributed generation is integrated with information on demand response and EVs.

[20] Providing a check on utility bills is a benefit that doesn't get much play when HEMS is discussed.

[21] HEMS tools and data feedback has brought the family through the cycle to prosumers.

[22] This is an example of aggregated thermal storage to provide DR during an extreme peaking event.

[23]   Small commercial customers are likely to be at the front of the line as energy consumers become empowered.

[24]   Information and control provide significant empowerment and reduce distractions.

[25]   Integrated ES, DR, and DG.

[26]   Clean energy behaviors will become a more powerful marketing tool to differentiate businesses as consumer empowerment grows and options for distributed energy proliferate.

[27]   A likely new service for utilities will be to provide aggregation services to consolidate load and encourage more collective behavior.

[28]   This is a good example of a virtuous BPI cycle described in Chapter 2.

[29]   Green tag and white tag markets will develop to pool renewable energy and energy efficiency credits.

Acronyms and Abbreviations

4C	comfort, convenience, cost, and carbon
3G	third generation wireless telecommunication network
4G	fourth generation wireless telecommunication network
AGC	automatic generation controller
AGI	advanced grid infrastructure
AMI	advanced meter infrastructure
AMR	automatic meter reading
APPA	American Public Power Association
ARRA	American Recovery and Reinvestment Act of 2009
AWEA	American Wind Energy Association
BEMS	building energy management systems
BPL	broadband over power line
CCET	Center for the Commercialization of Electric Technologies
CES	community energy storage
CFL	compact fluorescent lighting
CHP	combined heat and power
CIP	Common Internet Protocol
CO_2	carbon dioxide
CPP	critical peak pricing
CSAF	current state architectural framework
DA	distribution automation
DCS	distributed control system

DER	distributed energy resources
DG	distributed generation
DMS	distribution management system
DNP3	Distributed Network Protocol, version 3
DOE	Department of Energy
DPS	digital premise server
DR	demand response
DRCC	Demand Response Coordinating Committee
DRSGC	Demand Response Smart Grid Coalition
DVD	digital video disc
EDF	Environmental Defense Fund
EE	energy efficiency
EEI	Edison Electric Institute
eMUD	energy municipal utility district
EMS/SCADA	Energy Management Systems and Supervisory Control and Data Acquisition
EPRI	Electric Power Research Institute
ES	energy storage
ESA	Energy Storage Association
ERCOT	Electric Reliability Council of Texas
EV	electric vehicle
EVSE	electric vehicle support equipment
FDIR	fault detection, isolation, and restoration
FERC	Federal Energy Regulatory Commission
GIS	geospatial information system
GWA	GridWise Alliance
GWAC	GridWise Architecture Council
HEMS	home energy management system
HPPV	high penetration PV
HVAC systems	heating, ventilation, air conditioning systems
ICT	information and communication technology
IEC	International Electrotechnical Commission
IEEE	Institute of Electrical and Electronics Engineers
IHD	in-home displays

IOU	investor-owned utility
IP	Internet Protocol
ISO	independent system operator or International Standards Organization
ISSGC	Illinois Statewide Smart Grid Collaborative
IT	information technology
ITIL	Information Technology Infrastructure Library
ITT	information technology and telecommunications
KPI	key performance indicator
kW	kilowatt
kWh	kilowatt-hour
LCOE	levelized cost of energy
LED	light emitting diode
LEED	Leadership in Energy and Environmental Design
LP	long-playing vinyl record
LPPC	Large Public Power Council
LTE	long-term evolution
MCC	Microelectronics & Computer Technology Corporation
MDM	meter data management
MDMS	meter data management system
MDU	multidwelling unit
MOU	municipally owned utility
MUD	municipal utility district
MW	megawatt
MWh	megawatt-hour
ODM	original device manufacturer
OEM	original equipment manufacturer
OLA	operational level agreements
OMS	outage management system
NARUC	National Association of Regulatory Utility Commissioners
NASUCA	National Association of State Utility Consumer Advocates
NERC	North American Electric Reliability Corporation
NIST	National Institute of Standards and Technology
NOC	network operating center

NRECA	National Rural Electric Cooperative Association
NREL	National Renewable Energy Labs
NRTC	National Rural Telecommunications Council
PHEV	plug-in hybrid electric vehicle
P&L	profit and loss statement
PLC	power line carrier or programmable logic controllers
PMU	phasor management units
PSC	Public Service Commission
PUC	Public Utility Commission
PV	photovoltaic
RAP	Regulatory Assistance Project
RFI	request for information
ROI	return on investment
RPS	renewable portfolio standard
RTP	real-time pricing
RTU	remote terminal unit
RUP	rational unified process
SCADA	supervisory control and data acquisition
SEER	standard energy efficiency rate
SEIA	Solar Energy Industry Association
SEPA	Solar Electric Power Association
SGAF	Smart Grid Architecture Framework
SGCC	Smart Grid Consumer Coalition
SGIP	Smart Grid Interoperability Panel
SGOE	Smart Grid Optimization Engine
SLA	service level agreement
Solar PV	solar photovoltaic
TCO	total cost of ownership
TOU	time of use
TSW	true sine wave
USGBC	U.S. Green Building Council
UTC	Utilities Telecommunications Council
V2G	vehicle-to-grid
V2H	vehicle-to-home

VoIP	Voice over IP
VPP	virtual power plant
WiMAX	Worldwide Interoperability for Microwave Access

About the Authors

Andres Carvallo is a well-known visionary, technologist, speaker, author, and operations expert with 25 years of experience in the energy, wireless, computer, and software industries. Mr. Carvallo is a board member of the Utilities Telecom Council's Smart Networks Council and an energy and technology advisor to The University of Texas at Austin, The University of Kansas, and The University of California Los Angeles. Having held executive management positions since 1992, he has been responsible for the successful development and commercialization of more than 40 products, in companies ranging from start-ups to market leaders. Currently, Mr. Carvallo is the executive vice president and chief strategy officer at Proximetry, a leading wireless network and performance management provider. Prior to Proximetry, Mr. Carvallo was the executive vice president and chief strategy officer at Grid Net, a pioneer smart grid software platform provider. Prior to Grid Net, Mr. Carvallo was for 7 years the chief information officer and chief architect at Austin Energy, where he was responsible for the utility's technology vision, planning, development, and operations. While at Austin Energy, Mr. Carvallo defined and started promoting the term "smart grid" in 2004. He also invented the first smart grid fully deployed in the United States. Mr. Carvallo became a premier smart grid industry thought leader and led a comprehensive and highly lauded technology transformation company-wide. As the chief architect of the first successfully deployed smart grid in the United States, he has a unique prospective on the strategies, business cases, project management focus, architectures, best practices, technologies, and products that work and those that do not. Mr. Carvallo is a requested speaker on smart grid, CleanTech, green IT, SOA, wireless, and running IT as a business. His professional successes have been recognized with 22 industry awards, including 2005 IT Executive of the Year by the Association of Information Technology Professionals; the 2006 Premier 100 IT Leader and Best in Class of Premier 100 by *Computerworld;* the 2006 CIO 100 Award by *CIO Magazine;* 2007, 2008, 2009, and 2010 InformationWeek 500; 2008 and 2009 Top 12 Green IT Company by *Computerworld;* 2009 Computerworld Honors Laureate, Finalist, and 21st Century Achievement Award; 2009 CIO Hall of Fame Finalist by *CIO Magazine;* 2009 CIO of the Year by Energy Central; and 2009 and 2010 UtiliQ #2 Company by *Intelligent Utility* magazine and IDC. Prior to working at Austin Energy, Mr. Carvallo was a chief executive officer, executive vice president, and vice president at four other start-ups, a president and general manager at Philips Consumer Communications, a general manager at Digital Equipment Corporation, a general manager at Borland, a product manager at SCO, and a product manager for Windows at Microsoft. Mr. Carvallo received a B.S. in mechanical engineering from the University of Kansas with a concentration in robotics and control systems and he has completed executive management programs at the University of Idaho, Stanford University, and The Wharton School of the University of Pennsylvania.

John Cooper is a creative thinker, author, researcher, and project manager, active in the energy, telecommunications, IT services, and government research industries since the mid-1980s. With extensive experience in business development and consulting positions at innovative electric and telecommunications companies, Mr. Cooper has been responsible for leading innovation projects in all aspects of the emerging smart grid, ranging from utility IP networks, wireless advanced metering infrastructure, distributed generation, demand response, energy efficiency, utility-scale energy storage, virtual power plants, and electric vehicle charging infrastructure. Most recently, Mr. Cooper was the vice president for utility solutions at Grid Net, a pioneer smart grid software platform provider. Before Grid Net, Mr. Cooper provided consulting services as the president of his consulting firms Ecomergence and MetroNetIQ, where notable clients included Sharp Labs of America (research on virtual power plant business models), the Pecan Street Project (smart grid project manager and author of the technical report), Xtreme Power (utility-scale energy storage company), GRIDbot (electric vehicle charging start-up), various clients regarding multipurpose broadband networks for advanced metering infrastructure and other applications, and Austin Energy (pioneer GENie Project in 2004). In the mid-1990s, Mr. Cooper worked with Central and South West, an IOU now merged with AEP, to plan one of the first wireless AMR projects in Tulsa, Oklahoma. Early in his career, he helped launch and then served as the director of the Texas Senate Research Center. Mr. Cooper is the author of The ABCs of Community Broadband, a guide for community leaders, as well as numerous white papers and magazine articles on smart grid. Mr. Cooper received a B.A. in government from the University of Texas at Austin and an M.B.A. with honors from the McCombs School of Business at the University of Texas at Austin.

Index

Operational level agreements (OLA), 106
Organizations, *xx, xxv, xxvi*, 1, 10, 12, 19,
    20, 22, 23, 24, 37, 41, 48, 64, 68,
    71, 78, 83, 86, 99, 102, 103, 106,
    107, 108, 114, 115, 117, 118,
    121, 122, 124, 130, 131, 132,
    136, 139, 141, 142, 144, 148,
    149, 151, 152, 153, 154, 155,
    156, 157, 158, 159, 160, 161,
    162, 163, 164, 165, 166, 167,
    168, 169, 170, 171, 173, 174,
    175, 181, 182, 197, 198, 200,
    208, 213
Original device manufacturer (ODM), 168
Outage management system (OMS), 2, 18,
    29, 36, 40, 101, 188, 199

Phasor management units (PMU), 199
Photovoltaic (PV), 30, 60, 63, 67, 68, 69,
    70, 75, 138, 166, 176, 207
Plug-in hybrid electric vehicle (PHEV), 104,
    119
Profit and loss statement (P&L), 32
Programmable logic controllers (PLC), 9,
    29, 164
Public Service Commission (PSC), 160, 161
Public Utility Commission (PUC), 160, 161

Rational unified process (RUP), 108, 115
Real-time pricing (RTP), 74, 117, 133, 143,
    196
Regulatory Assistance Project (RAP), 161
Remote terminal unit (RTU), 8, 26, 27, 29
Renewable energy, 154, 155, 174, 175
Renewable portfolio standard (RPS), 154,
    155, 174, 175
Request for information (RFI), 160, 165
Return on investment (ROI), 103, 104

Service level agreement (SLA), 36, 52, 106,
    116
Smart grid, *xv, xvi, xvii, xix, xx, xxi, xxii,
    xxiii, xxiv, xxv, xxvi, xxvii*, 1, 2, 3,
    4, 6, 7, 8, 11, 13, 14, 16, 17, 18,
    19, 21, 22, 23, 24, 25, 27, 28, 29,
    30, 31, 35, 36, 37, 38, 39, 40, 41,
    42, 43, 44, 45, 46, 47, 48, 49, 50,
    51, 52, 55, 57, 58, 60, 63, 64, 65,
    67, 68, 70, 74, 78, 79, 80, 83, 84,
    86, 87, 89, 90, 91, 92, 93, 94, 96,
    97, 98, 99, 101, 102, 104, 105,
    106, 107, 108, 109, 110, 111,
    112, 113, 115, 116, 117, 118,
    149, 120, 121, 122, 124, 125,
    126, 127, 128, 129, 131, 132,
    133, 136, 137, 138, 139, 141,
    142, 144, 145, 147, 148, 149,
    150, 151, 152, 153, 154, 156,
    157, 158, 159, 160, 161, 162,
    163, 164, 165, 166, 167, 168,
    169, 170, 171, 172, 173, 174,
    175, 176, 177, 178, 183, 184,
    185, 186, 187, 188, 189, 190,
    192, 193, 194, 196, 197, 198,
    199, 200, 202, 203, 205, 206,
    208, 212, 213, 218, 219, 220,
    221, 222, 223
Smart Grid 1.0, *xxii, xxv*, 2, 83, 85, 87, 89,
    91, 97, 99, 101, 103, 105, 107,
    109, 111, 113, 115, 117, 120,
    121, 129, 202
Smart Grid 2.0, *xxii, xxv*, 2, 117, 119, 120,
    121, 123, 125, 126, 127, 159,
    130, 131, 133, 135, 137, 139,
    141, 143, 145, 202
Smart Grid 3.0, *xxvii*, 2, 183, 184, 185, 187,
    189, 191, 193, 195, 197, 199,
    201, 203, 205, 206, 207, 209,
    211, 213, 215, 217, 219, 220,
    221, 223
Smart Grid Architecture Framework
    (SGAF), 197, 198
Smart Grid Consumer Collaborative
    (SGCC), 165, 175
Smart Grid Interoperability Panel (SGIP),
    22, 24, 151, 162, 163, 174
Smart Grid Optimization Engine (SGOE),
    38, 40, 41, 42, 44, 197, 200, 201,
    202, 213, 218, 222
Solar Electric Power Association (SEPA), 166
Solar Energy Industry Association (SEIA),
    168
Solar photovoltaic (solar PV), 15, 28, 30, 36,
    47, 59, 60, 63, 67, 68, 69, 70, 75,
    78, 105, 119, 120, 128, 134, 137,
    138, 140, 141, 155, 158, 166,
    168, 169, 176, 202, 207, 211,
    212, 223
Standard energy efficiency rate (SEER), 64
Supervisory control and data acquisition
    (SCADA), 2, 8, 31, 38, 39, 48,